POWER DISTRIBUTION NETWORK DESIGN FOR VLSI

POWER DISTRIBUTION NETWORK DESIGN FOR VLSI

QING K. ZHU
Intel Corporation
Matrix Semiconductor Inc., U.S.A.

WILEY-INTERSCIENCE

A JOHN WILEY & SONS, INC., PUBLICATION

For general information on our other products and services please contact our Customer Care Department within the U.S. at 877-762-2974, outside the U.S. at 317-572-3993 or fax 317-572-4002.

Wiley also publishes its books in a variety of electronic formats. Some content that appears in print, however, may not be available in electronic format.

Library of Congress Cataloging-in-Publication is available.

ISBN 0-471-65720-4

Printed in the United States of America.

10 9 8 7 6 5 4 3 2 1

CONTENTS

PREFACE

This book provides the detailed information on power distribution network design in integrated circuit chips. Power distribution network design is a critical part of the job in circuit design and physical integration for high-speed chips.

The *IR* drop and *di/dt* noise associated with the power distribution networks are crucial to circuit timing and performance. Due to the complexity of the millions of gates and interconnects in modern VLSI chips, power network analysis is accomplished using CAD tools. These tools take the layout database, usually in GDSII files, extract the RC parasitic for the power distribution network, and model the current consumption for switching devices.

A fast circuit simulation is done for the electrical model of the power distribution network in order to determine the *IR* drop or other supply voltage noises, as well as the current density of metal power lines for checking electromigration failures.

In addition, the decoupling capacitors are inserted into the power network for stabilizing the supply voltages in local regions where current surges occur from time to time due to clock and logic operations. The decoupling capacitors and power distribution networks are required in some optimal form not only on-chip, but also on the package and at system levels.

This book will explain the design issues, guidelines, examples, and CAD tools for the power distribution of the VLSI chip and package. The user guide of the VoltageStorm™ tool from Cadence Design Systems, Inc. is referred to throughout [51], together with the author's experience using this tool in designs.

The book is organized into seven chapters. Chapter 1 is an introduction to the power supply network, power network modeling, decoupling capacitors, and process scaling trends. Chapter 2 illustrates the design perspectives for the power distribution network, including power network planning, layout specifications, decoupling capacitance insertion, modeling and analysis of power networks, and IR drop analysis and reduction. Chapter 3 explores electromigration phenomena for the on-chip power distribution network.

Chapter 4 discusses IR drop analysis methodology. It is taken primarily from the VoltageStorm™ tool, using both static and dynamic analysis methods. The static method is performed for some level worst-case IR drop analysis without the knowledge of input vectors at the chip's primary inputs. Chapter 5 describes the commands and user interfaces of the VoltageStorm™ tool from Cadence Design Systems, Inc. [51]. Chapter 6 lists the microprocessor design examples, with a focus on on-chip power distribution. Readers will gain the insights into industry chip design for power distribution networks from these examples.

Chapter 7 discusses the flip-chip and package design issues, since the package is a part of the global power distribution. A case study has been provided in this chapter for selecting the package options, based on the performance requirements for the power supply. Power network measurement techniques from silicon are also discussed at the end of Chapter 7.

A glossary of key words and basic terms is provided at the end of the book to help understand the basic concepts in VLSI design and power distribution.

With the continually decreasing supply voltages and the increasing transistor switching currents on-chip, power supply noises on-chip remains the challenging issue for high-performance chip design. More and more research will be needed in the future in CAD tools for switching current modeling and accurate power network analysis. The design methodology for power delivery will need to consider the performance, layout area, and package technology optimization for future chips.

The author would like to thank Mr. George J. Telecki at John

Wiley & Sons, Inc. for providing the chance to get this book published. He also thanks his co-workers in Intel Corporation, including David Ayers, Alex Waizman, and Bendik Kleveland. Finally, he appreciates the strong support from family members, including wife Huiling Song and two sons Phillip and Michael.

1

INTRODUCTION

As power supply voltage continues to drop with the VLSI technology scaling associated with significantly increasing device numbers in a die, power network design becomes a very challenging task for a chip with millions of transistors. The common task in VLSI power network design is to provide enough power lines across the chip to reduce the voltage drops from the power pads to the center of the chip. The voltage drops are mainly caused by the resistance or inductance of the power network metal lines.

The power network can be modeled as a low-pass filter with RL segments in series, attached with capacitors at each end. The current sources of the switching gates and the intentional decoupling capacitors are also inserted in the model. The IR drop is proportional to the average current consumed by the circuit in the chip. The $L \cdot di/dt$ drop is proportional to the time-domain change of the current, due to the switching of logic gates in the chip operations.

This chapter is organized into seven sections. Section 1.1 discusses the general trend of power supply noise with the process technology scaling. Section 1.2 shows the modeling methodology for on-chip power networks. Section 1.3 discusses the switching current modeling methodology for the power distribution network, which is critical for the accuracy of power grid analysis. Once we obtain the models, the power network can be characterized as a linear network with R, L, C, and current sources, in order to solve the voltage distributions across the power network.

Power Distribution Network Design for VLSI, by Qing K. Zhu
ISBN 0-471-65720-4 © 2004 John Wiley & Sons, Inc.

Section 1.4 discusses a special topic in power network design: the decoupling capacitor optimization to allocate enough decoupling capacitors between V_{dd} and V_{ss} nets, but not over-allocating so as to result in enlargment of the die area. Section 1.5 discusses the on-chip inductance effects on power network modeling. We show the metal configurations used in the power line design in order to minimize the inductance delay. In general, many thin-width V_{dd} and V_{ss} lines interleaved with each other in the power distribution network are preferred in order to minimize the area of the return current loop or on-chip inductance.

Section 1.6 discusses process technology scaling impacts for the future power network design. We discuss the technology scaling impacts in two scenarios. Section 1.7 provides the summary to this chapter.

1.1   POWER SUPPLY NOISE

Noise problems in microprocessor power distribution networks have been discussed in the literature [1, 2, 3, 4, 5, 6]. The supply voltage is continually dropping in microprocessor design to reduce the power consumption and matche the reduced gate oxide thickness in the scaled IC process technology generations. Figure 1-1(a) shows the supply voltage drop trend in new technologies; and Figure 1-1(b) shows the gate oxide thickness reduction during the process scaling.

The on-chip decoupling capacitor is constructed by using the dummy transistors connected to V_{cc} with the gate, and V_{ss} with the drain and source. A conventional method for on-chip decoupling capacitance allocation is based on a percentage (i.e., 10%) area in each layout window (e.g., 100 × 100 μm) allocated for the decoupling capacitance.

The decoupling capacitors are inserted near the large-size buffers, such as clock buffers or phase-locked loops. The conventional method, based on the layout area percentage, is not optimal, either being overestimated for a large layout area or underestimated for meeting the power noise requirements.

The power distribution design techniques used for DEC Alpha chips, such as the C4 package and on-chip power planes, can be found in [1]. The decoupling capacitance optimization technique, based on the layout floor plan graph and path-finding algorithm,

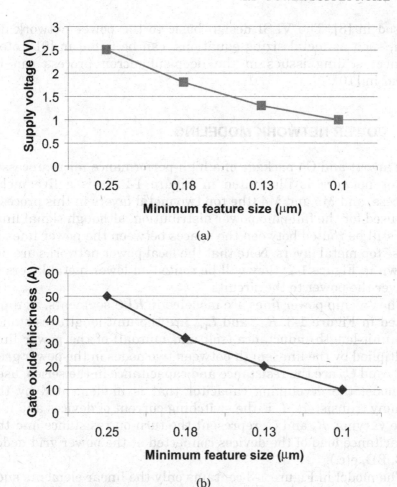

Figure 1-1. Power supply (a) and gate oxide scaling (b) trends.

can be found in [2]. The power network modeling and analysis techniques for PowerPC microprocessors can be found in [3]. A power network modeling and simulation CAD tool is described in [4].

The reliability problems (i.e., electromigration) and CAD tool for the power network are discussed in [5]. The basics of VLSI power distribution can be found in [6]. The description of a high-performance power network scaling model and decoupling capacitance optimization method is proposed in [7]. A criterion to include the inductance in on-chip interconnect modeling was discussed in [8]. The VLSI design basic to the power network design, such as metal sizing equations, can be found in [9]. Interconnect scaling issues in the deep-submicron process can be found in [10].

1.2   POWER NETWORK MODELING

The layout and C4 package of a high-performance microprocessor power network is illustrated in Figure 1-2. It is a five metal process, and M5 and M4 (the top two metal layers in this process) are used for the full-chip power distribution, although signal lines can still be routed between the spaces between the power lines in these top metal layers. Note that the local power networks are not shown in Figure 1-2; they will be routed on lower metal layers to deliver the power to the circuits.

The on-chip power lines are modeled in RLC segments, as illustrated in Figure 1-3. R_{vcc} and L_{vcc} are the unit-length resistance and unit-length inductance (self and mutual) of the power line, multiplied by the line length between two nodes in the power grid.

R_d and C_d are the resistance and capacitance in the series, used to model the decoupling capacitor that is implemented by the dummy transistors. I_s is the switching current of devices and it is time varying. R_s and C_s represent the turn-on resistance and the capacitance load of the devices connected at the power grid nodes (AC, BD, etc.).

The model in Figure 1-3 contains only the linear elements such as R, L, C, and current sources. It suggests to us that a linear circuit simulator can be used to speed up the large-size microprocessor power network analysis based on the proposed model. The key parameters of decoupling capacitors (dummy transistors) are C_{decap} and R_{decap}, as shown in Figure 1-4.

The charges in C_{decap} are used to help the supply voltage stability in C_{sw} (switching gates) before the charges eventually come from the supply voltage source via the long current loop from the package.

To improve the efficiency of the decoupling capacitors, the R_{decap} needs to be sufficiently small. When V_{cc} is applied to the gate, as shown in Figure 1-5, the inversion channel is created between the D and S with the $R_{ds\text{-}on}$ resistance. The $R_{ds\text{-}on}$ resistance is the 1/slope of the I/V curves of the resistor at V_{ds} = 0V. The

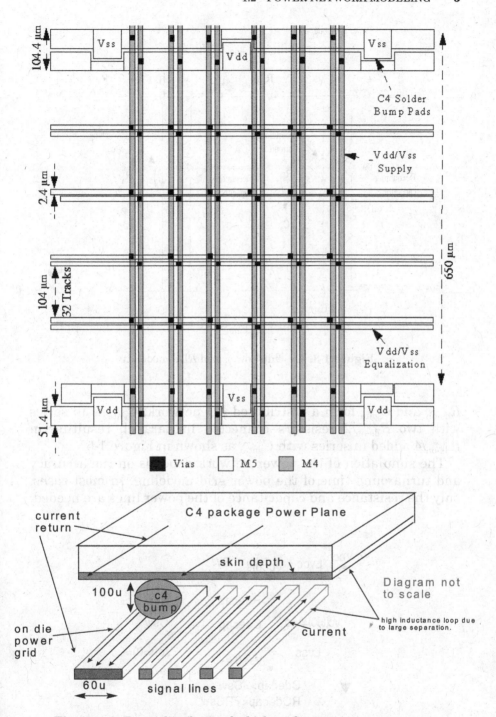

Figure 1-2. Power distribution for high-performance microprocessors.

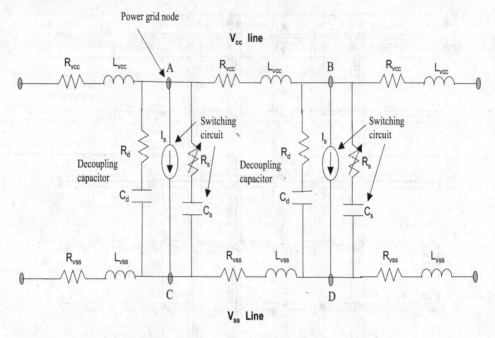

Figure 1-3. On-chip power grid RLC modeling.

$R_{\text{ds-on}}$ and C_{gate} form a distributed RC network. C_{gate} is in series with two $R_{\text{ds-on}}/2$ resistors connected in parallel, resulting in $R_{\text{ds-on}}/4$ added in series with C_{gate}, as shown in Figure 1-5.

The simulation of the power network depends on the accuracy and turnaround time of the power grid modeling. In most cases, only the resistance and capacitance of the power lines are needed,

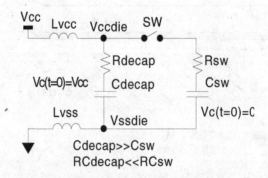

Figure 1-4. Switching model of decoupling capacitor.

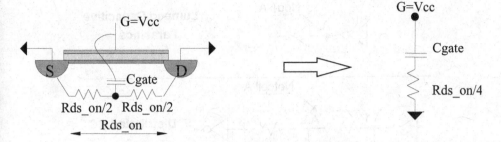

Figure 1-5. Decoupling capacitor modeling.

excluding the metal inductances for the on-chip power network. Many CAD tools are available for the purpose of extracting the interconnect RC for power grids, as summarized in Table 1-1.

The on-chip inductance for the power grid can be ignored by using special design rules, shortening the return loop of V_{dd} and V_{ss} by using several interleaved V_{ss} and V_{dd} lines, as shown in Figure 1-2, for example, to implement the power grid.

During the RC modeling process, each metal segment can be represented in two forms as follows: (1) the lumped capacitive parasitic, or (2) the distributed RC parasitic, as shown in Figure 1-6(a). The lumped capacitive parasitic represents the total wire capacitance from each driver circuit in the signal net. The distributed RC parasitic includes the resistance (R) of the metal line in the modeling.

Power grid modeling usually uses the RC model, since the metal line resistance of the power grid is significant at the full-chip level. A long metal line can be broken into multiple RC segments, as shown in Figure 1-6(b).

Table 1-1. Well-known RC extraction CAD tools

Tool	Manufacturer
Fire & Ice	Cadence Design Systems
Star-RCXT	Synopsys
xCalibre	Mentor Graphics
HyperExtract	Cadence Design Systems
Arcadia	Synopsys
Columbus	Sequence Design
Nautilus	Cadence Design Systems
QuickCap	Random Logic

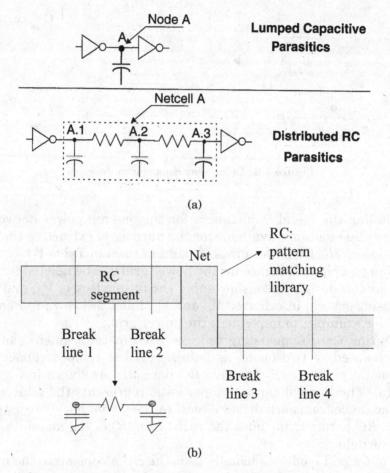

(a)

(b)

Figure 1-6. Lumped and distributed *RC* models.

Each *RC* segment is modeled with a series resistor, together with two capacitors at two ends of the resistor. The metal segment capacitance is evenly divided by two capacitors. This is usually called the Pai–*RC* model since it looks like a pi (π) symbol, as shown in Figure 1-6(b).

The extracted *RC* data from the layout are saved in a standard parasitic format (SPF) file. It includes a list of nets and detailed *RC* values. The *R* and *C* elements with the node names are specified either as schematic-based labels or layout-based labels, depending on the options used in the *RC* netlisting stage.

The schematic node names are preferred in the SPF, since this SPF can be back-annotated to the prelayout schematic netlist [33,

34]. In addition, the SPF can include the device section that models the extracted devices from the physical layout.

In general, the capacitance can be formed between any polygons in the layout, although the closer ones have more significant capacitances, and thus have more impact on the total capacitance of the net. Figure 1-7 shows the possible capacitances between the gates and metal lines in the physical layouts.

The capacitance to the substrate is dominant over other coupling capacitances in the old one or two metals technology. But the situation changes in the latest submicron technology with seven to eight metal layers, since the top-level metals are far away from the substrate, and the total capacitance of these top-level metals is more impacted by the coupling capacitances between adjacent lines in the same layer or adjacent layers of the layout.

In addition, the spacing between metal lines is continually scaled, so the coupling capacitance between neighboring metal lines becomes more and more important. The calculation of the resistance or capacitance can be done through the direct solution of the well-known Maxwell's EM equations or Green's functions [17].

A complex geometrical layout can require an extremely long computational time using the direct EM field solution. Therefore,

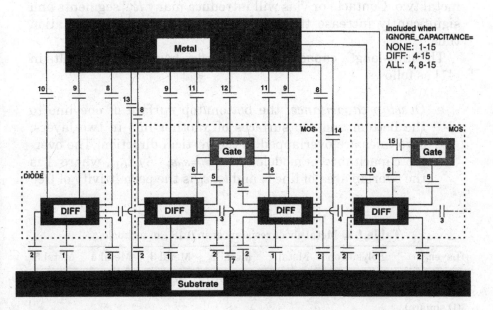

Figure 1-7. Coupling capacitances between conductors in a VLSI layout [33].

equations or capacitance models are usually adopted in the capacitance calculation for a large-scale layout.

Once the capacitance equations have been established, they are used in the RC extraction, which is fast enough to handle a large-scale layout. The RC extraction works on the physical database together with the specified RC equations.

Let us review the basic resistance equation:

$$R = sl/w \text{ (ohm)} \tag{1-1}$$

In Equation (1-1), s is the sheet resistance in the unit of ohm/square, l is the length of the line in μm, and w is the width of the line in μm.

Table 1-2 shows the sheet resistance data in a 0.18 μm technology. Metal four and metal five have significantly lower resistances, making them suitable for long metal routes. The polysilicon and metal one layers have high resistance, making them suitable for short metal connects.

The contacts or vias between metal layers, as shown in Figure 1-8, are usually modeled as resistors. Each contact or via has a fixed resistance based on design rules. The contact represents the metal hole between metal one to the diffusion or poly layer, whereas the via represents the metal hole between metal one and metal two. Contacts or vias will introduce many RC segments and significantly increase the RC parasitic file size and simulation time.

The unit-length capacitance models are based on the results in [41] as follows.

a. *Overlap capacitance:* the bottom/top surface of one line to the bottom and top surfaces of another line in two layers. Two lines are overlapped in the vertical direction. The overlap capacitance is modeled as $Ca = \varepsilon_0\varepsilon_r \cdot A/d_{l1l2}$, where A is the overlap area of line l_1 and l_2, ε_0 is the permittivity of free

Table 1-2. Metal sheet resistances in 0.18 μm technology

Layer	Polysilicon	Metal 1	Metal 2	Metal 3	Metal 4	Metal 5
Sheet Resistance (Ω square)	5.5	0.1	0.05	0.05	0.01	0.01

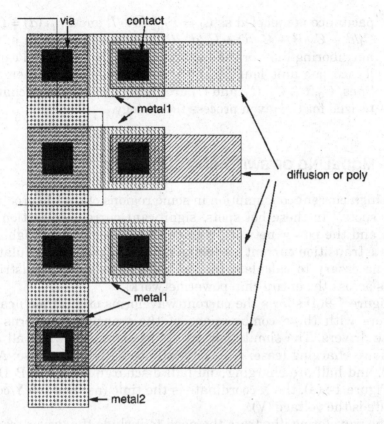

Figure 1-8. Contacts and vias [9].

space ($8.854 \cdot 10^{-14}$ F/cm^2), ε_r is the relative permittivity between l_1 and l_2, and d_{l1l2} is the vertical spacing between two lines.

b. *Fringe capacitance:* the side surface of one line to the bottom or top surface of another line in two layers. Two lines may or may not be overlapped in the vertical direction. The fringe capacitance is modeled as $C_{fr} = C_{fr0} \cdot l \cdot (e^{-x_1/x_0} - e^{-x_2/x_0})$. x_1 is the distance from l_1 (side edge) to l_2 (near-end edge), and x_2 is the distance to l_2 (far-end edge). l is the length of l_1 (side edge). C_{fr0} and x_0 are model coefficients that are characterized based on different vertical profiles. In a special case, two side edges may coincide in l_1 and l_2 ($x_1 = 0$ and $x_2 =$ width of l_2) and the model becomes $C_{fr} = C_{fr0} \cdot l \cdot (1 - e^{-x_2/x_0})$.

c. *Lateral capacitance:* the side surface of one line to the side surface of the adjacent line in the same layer. The lateral capacitance is modeled as $C_{lt} = F_{l1l2}(d) \cdot l$, and $F_{l1l2}(d) = C_0 + C_1/d + C_2/d^2 + C_3/d^3 + C_4/d^4$. l is the parallel length of two neighboring lines or conductors, $F_{l1l2}(d)$ is the lateral capacitance per unit length, and d is the spacing between two lines. C_0, C_1, C_2, C_3, and C_4 are coefficients that are characterized for the given process technology.

1.3   MODELING OF SWITCHING CURRENTS

The high current consumption in some regions of the die produces "hot spots." In these hot spots, significant current transition occurs and the power network voltage fluctuation will be high. Accurate transition current modeling and power network simulation are necessary to calculate the noise and temperature distributions across the entire chip power network.

Figure 1-9(a) shows the current waveforms of multiple nearby drivers with three combinations of the transition patterns for these drivers. The simulation results are obtained when all drivers are charging (case: ALL UP), all on discharging (case: ALL DN), and half are charging and half discharging (case: UP_DN). In Figure 1-9(a), the X-coordinate is the time (ns) and the Y-coordinate is the voltage (V).

The waveforms illustrate the need to include the driver transition patterns (UP/DOWN) to model the transition currents. In our simulation, a 295.2 μm long bus with 130 signals is simulated in the minimum M5 width and pitch. Figure 1-9(b) shows the circuit schematic to be simulated. Figure 1-9(c) shows the entire power grid modeling for the simulation. Figure 1-9(d) shows the structure of bus lines and V_{cc}/V_{ss} lines on the M5 layer included in the simulation.

In general, the total current consumption $I(t)$ of the CMOS circuit shown in Figure 1-10 consists of three components: I_d, I_{sc}, and I_l. I_d is the charge or discharge current to the output load:

$$I_d = C_{load}V_{cc}f \qquad (1-2)$$

In Equation (1-2), C_{load} is the total output load of the driver, including the gate load and interconnect load; V_{cc} is the supply voltage; and f is the switching activity of C_{load}. Although the charge and discharge dynamic current I_d is a predominant component of the total current consumption, other two current components (I_{sc}, I_l) are still significant in the submicron CMOS process.

The short-circuit current I_{sc} is due to the fact that pMOS and nMOS transistors are both in the transition region of the inverter. The leakage current I_l is due to the reverse-biased diode's leakage between the diffusion region and the substrate or well. Although the sum of the short-circuit and leakage currents accounts for less than 15% of the total current consumption of the microprocessor chip, the percentage will go up in future CMOS processes.

Figure 1-10(b) shows the current waveforms based on the estimated current components; the waveform is assumed to be a tri-

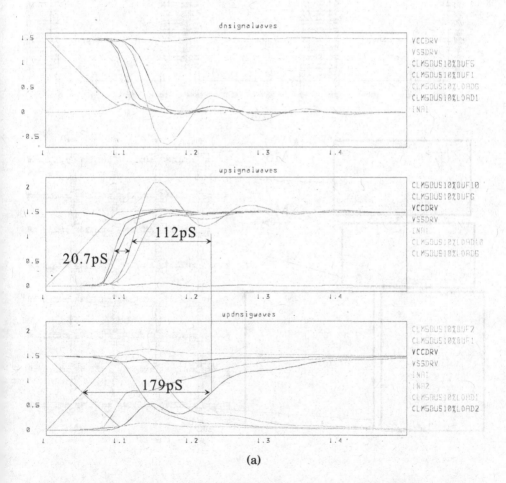

(a)

Figure 1-9. Switching noise simulation based on power grid modelling. (a) Simulation result. (*Figure continues on next page*)

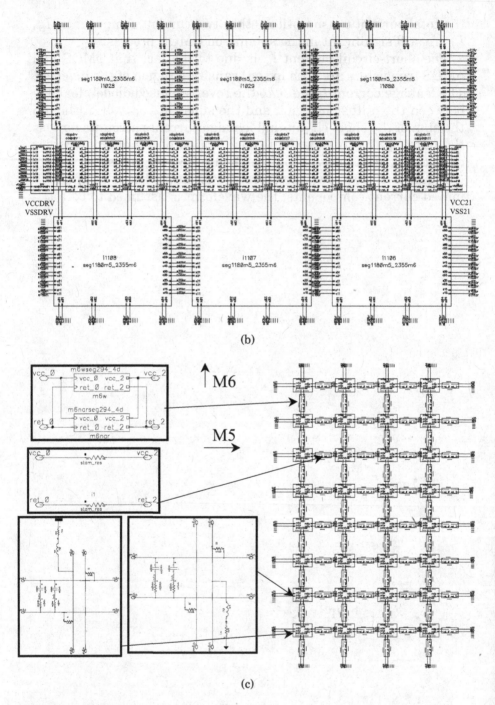

(b)

(c)

Figure 1-9 *(continued).* (b) Simulated circuit. (c) M5 and M6 power grid modelling. *(Figure continues on next page)*

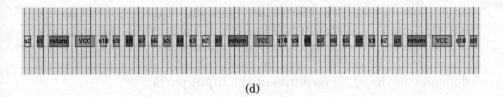

(d)

Figure 1-9 *(continued).* (d) Bus lines layout structure.

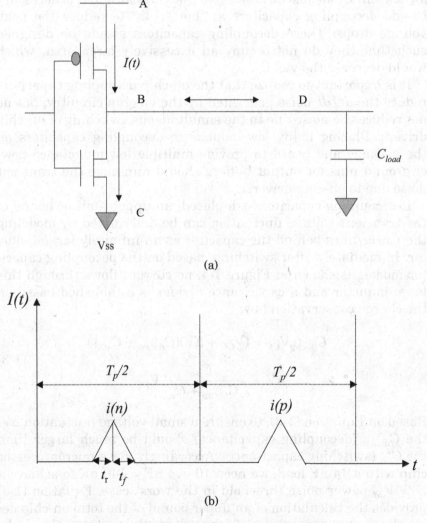

Figure 1-10. Modeling of switching currents.

angle. The current waveforms are back-annotated into the power network model, as shown in Figure 1-3. To improve the accuracy of the current waveforms, a current simulation tool such as Synopsys, Inc.'s PowerMill™ can be used, although the result largely depends on the (0, 1) patterns at the input ports.

1.4   ON-CHIP DECOUPLING CAPACITANCE

To prevent the supply level from collapsing when many gates switch simultaneously at the same clock transition, it is necessary to add decoupling capacitors at "hot spots" to reduce the peak voltage drops. These decoupling capacitors should be designed such that they do not occupy an excessively large area, which would decrease the yield.

It is important to realize that the on-chip decoupling capacitors reduce the di/dt noise generated by the on-chip circuitry, but do not reduce the noise due to the simultaneous switching of off-chip drivers. Placing many low-inductance decoupling capacitors on the package and board to provide multiple low-inductance power/ground pins for output buffers should minimize the transient noise due to off-chip drivers.

If decoupling capacitors are placed, an upper limit or bound of the transient voltage fluctuation can be determined by modeling the power lines behind the capacitor as an infinitely large inductor. Immediately after switching, based on the decoupling capacitor model, as shown in Figure 1-4, no current flows through this large inductor and a capacitance divider is established based on the charge conservation law:

$$C_{\text{decap}}V_{CC} = (V_{CC} + \Delta V)(C_{\text{decap}} + C_{sw})$$

$$\Delta V = -\frac{C_{sw}}{C_{\text{decap}} + C_{sw}}V_{CC}$$

(1-3)

Based on Equation (1-3), to ensure a small voltage fluctuation ΔV, the C_{decap} (decoupling capacitance) should be much larger than the C_{sw} (switching capacitance). Accordingly, for a microprocessor chip with a 14 nF load, we need $10 \cdot 14$ nF = 140 nF to achieve a 10% V_{dd} power noise threshold in the worst case. Equation (1-3) provides the calculation of an upper bound of the total on-chip decoupling capacitance to satisfy the voltage fluctuation ΔV bound.

The objective of the decoupling capacitance optimization problem is to minimize the total amount of decoupling capacitance as needed. Meanwhile, all the nodes in the power network model are satisfied with the specified supply voltage noise thresholds. Formally, we can describe the objective and constraints as follows [83]:

$$\text{Min} \sum_{n_i} (C_d)_i \qquad \text{Subject to } V_1 \leq V(n_i) \leq V_2 \qquad (1\text{-}4)$$

In Equation (1-4), $(C_d)_i$ is the decoupling capacitance and $V(n_i)$ the voltage at node n_i of the power network model, as shown in Figure 1-3; V_1 and V_2 are the lower and upper thresholds required for feasible supply voltages. We define a *noisy node* in the power network model as one in which, at some time, the voltage exceeds the required $[V_1, V_2]$ thresholds, as shown in Figure 1-11.

The thresholds are at the upper bound and lower bound away from the nominal supply voltages to guarantee the correct circuit timing. For example, with a nominal voltage of 1.3 V and 10% away allowed, the upper and lower thresholds are $[V_1, V_2] = [1.17$ V, 1.43 V].

The power network, with each node's transient voltages in the electrical model satisfying the given thresholds, is called a *feasible power network*. Adding the decoupling capacitors at noisy nodes will turn a power network into a feasible one. Figure 1-12(a) shows

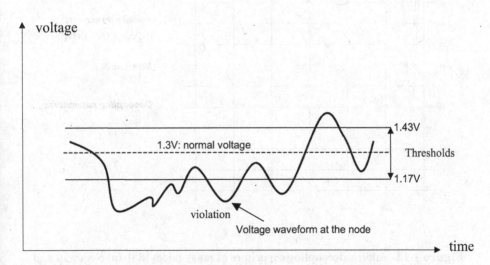

Figure 1-11. Supply voltage thresholds and noisy nodes definition [83].

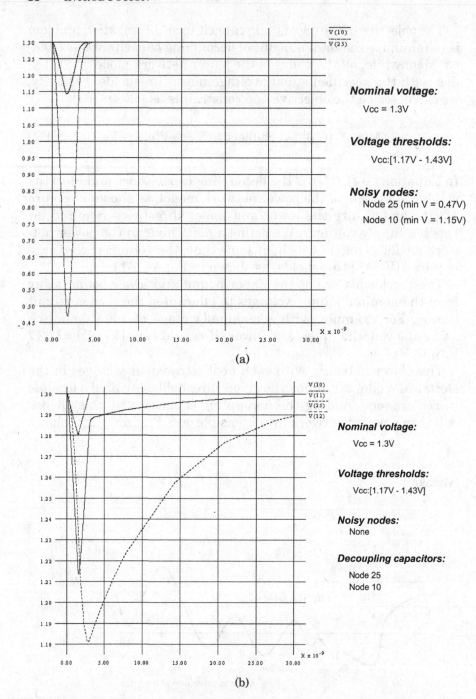

Figure 1-12. Adding decoupling capacitors at noisy nodes [83]. (a) Nodes 10 and 25 are noisy. (b) Adding more capacitors on Nodes 10 and 25.

one example with the simulated voltages of two nodes (Node 25 and Node 10) in the power network.

The minimum voltages (0.47 V and 1.15 V) of these nodes are less than the required lower threshold (1.17 V), and thus they are noisy nodes. The decoupling capacitor is added at each of these two noisy nodes and the voltages eventually satisfy the required thresholds, as shown in Figure 1-12(b).

Figure 1-13 shows the high-level decoupling capacitance optimization flow [83]. Procedure I adds the decoupling capacitors at the noisy nodes. Procedure II removes the unnecessary decoupling capacitance overallocated initially.

We have done experiments on a power network model with about 100 *RLC* grids and decoupling capacitors. Current sources have been added at each node in the model for transistor transitions with the current waveforms, as shown in Figure 1-10(b). The

Procedure I: Decoupling Capacitance Increment

Simulate the power network model with *RLC* elements and current sources.
Identify the "noisy" nodes by comparing the voltage results with the specified thresholds.
While (there is "noisy" node){
    For (each "noisy" node){
        Add a step size of the decoupling capacitance.
    }
    Simulate the power network model with the updated decoupling capacitance.
    Identify "noisy" nodes by comparing simulation voltages with the required thresholds.
}

Procedure II: Decoupling Capacitance Decrement

For (each node){
    Mark the node as "deductible";
}
While (there is still "deductible" node){
    Deduct a step size of decoupling capacitance from each "deductible" node;
    Simulate the power network model with the updated decoupling capacitance;
    Identify the "noisy" nodes by comparing simulation voltages with the required thresholds;
    For (each "noisy" node){
        Add a step size of the decoupling capacitance;
        Make the node as "nondeductible";
    }
}

Figure 1-13. Decoupling capacitance optimization flow [83].

cycle time is 3 ns or 330 MHz frequency in the experiments. Two voltage sources are added to model the C4 package power pads. The RL parasitic (200 Ω and 0.5 nH) of the package layer are included in the model. The nominal supply voltage is 1.3 V.

The power grid simulation is done using a fast linear circuit simulator [20]. The flow shown in Figure 1-13 is used to determine the locations and amounts of on-chip decoupling capacitors. Figure 1-14 shows the experimental results for a sensitivity study to decoupling capacitances. The decoupling capacitance is most sensitive to the changes in the noise margin and device transition currents.

This suggests to us that the model of the current consumption is the key to getting the accurate voltage drop and decoupling capacitance amounts. In addition, we want to reduce the on-chip decoupling capacitance size by improving the noise margin. This can be achieved by improving the power distribution on the package and the board. The changes of power line RLC values, as well as the absolute supply voltages with the same noise thresholds, do not show significant impact on the decoupling capacitance.

In the experiment, we assigned the initial RLC values at each node of the power network as follows: $R = 40\ \Omega$, $L = 0.005$ nH, $C = 0.3$ pF (without the decoupling capacitance at this initial assignment). The change of on-chip power line inductance does not lead to a lot of variation in decoupling capacitance, as shown in Figure 1-14(b); this is due to the very small L/R delay (0.12 ps) compared to the RC delay (12 ps) in this example.

The decoupling capacitor can be improved by using either the PN junction or a MOS varactor device [43]. As shown in Figure 1-15(a), the PN junction is formed by diffusing p+ doping in an n-well. As shown in Figure 1-15(b), the MOS varactor is formed by placing an nMOS in an n-well. The n-well is added to form a channel between the source and drain. In addition, V_{tune} and V_{gate} voltages are controlled to vary the gate capacitance used for the decoupling capacitances between V_{dd} and V_{ss}.

1.5   ON-CHIP INDUCTANCE

The inductive drop or noise ($L \cdot di/dt$) on the power lines becomes significant for high-speed microprocessor chips [14, 15], especially

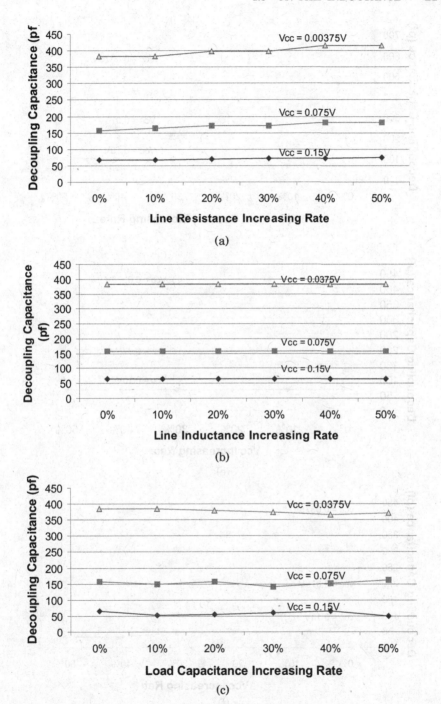

Figure 1-14. Sensitivity study of on-chip decoupling capacitances [83]. *(Figure continues on next page)*

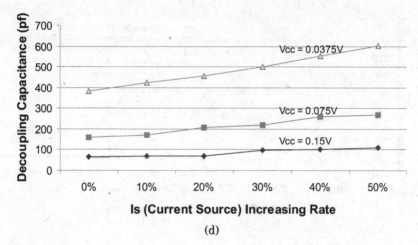

Is (Current Source) Increasing Rate

(d)

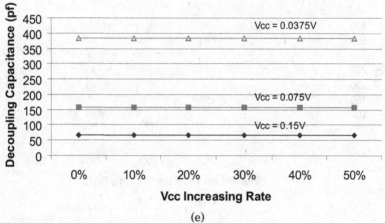

Vcc Increasing Rate

(e)

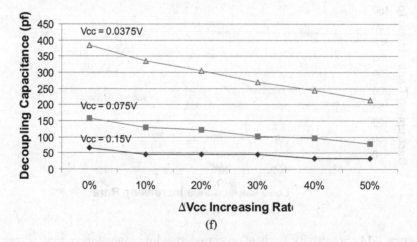

ΔVcc Increasing Rate

(f)

Figure 1-14 (*continued*).

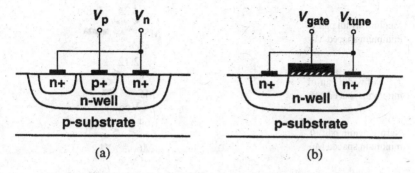

Figure 1-15. Decoupling capacitor [43]. (a) PN Junction. (b) MOS varactor.

when the chip becomes faster and larger in size. The characteristic impedance is $Z_0 = \sqrt{L/C}$. Adding decoupling capacitors will increase the capacitance but does not affect the inductance of the power planes. As a result, Z_0 is reduced, and current spikes generate smaller voltage drops because $\Delta V = Z_0 \Delta I$

Low impedance of the power network helps the pulse response and curbs the instantaneous fluctuations. The impedance Z_0 can be further reduced by lowering the inductance L of the power network. This section presents a metal wire design method to reduce the inductance by carefully selecting the sizes and spaces of power lines.

Figure 1-16(a) shows five different combinations of the widths and spaces for two adjacent V_{cc} and V_{ss} lines [21]. The inductance and resistance of these five combinations are shown in Figure 1-16(b) and Figure 1-16(c) for 10,000 μm long power lines. The inductance is calculated by using a two-dimensional model with the current loops between adjacent V_{ss} and V_{cc} lines. The first-order estimation of the unit-length loop inductance for two adjacent V_{cc} and V_{ss} lines is as follows:

$$L = \mu \frac{s}{w} \qquad (1\text{-}5)$$

In Equation (1-5), μ is the permeability of the dielectric material between adjacent V_{cc} and V_{ss} lines, s the space between the V_{cc} and V_{ss} lines, and w the width of V_{cc} or V_{ss} lines. The V_{cc} and V_{dd} nets are interchangeable in this book. Usually, V_{cc} is used for the analog signal and V_{dd} for digital design.

The inductance becomes large when the line space is big, which

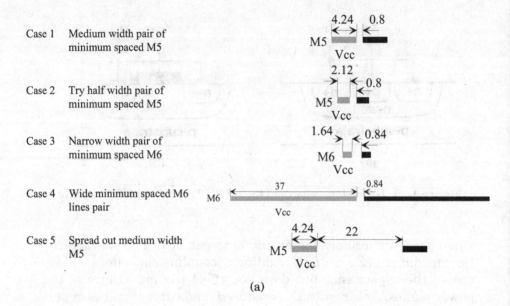

(a)

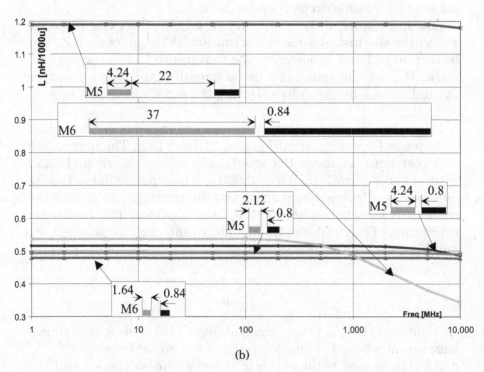

(b)

Figure 1-16. Characterization results of V_{dd}/V_{ss} metal structures [21]. (a) V_{cc} and V_{ss} cases. (b) On-chip inductance characterizations.

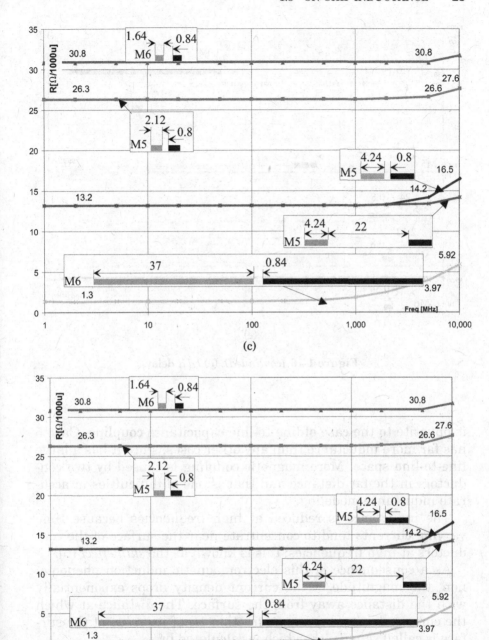

Figure 1-16 *(continued).* (c) Resistance characterizations. (d) Impedance calculation. *(Continued on next page)*

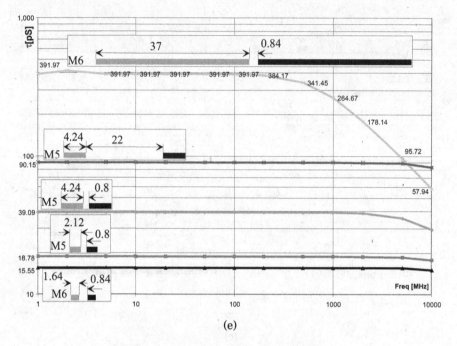

(e)

Figure 1-16 *(continued)*. (e) L/R delay.

is opposite to the case of line-to-line capacitance coupling. Case 5 has far more inductance than any other cases, since it has a large line-to-line space. More magnetic coupling is caused by two conductors in the far distance and that is one of difficulties in accurate inductance modeling.

The inductance is reduced at high frequencies because time varying currents tend to concentrate near the surface of the conductors at high frequencies; this is known as the *skin effect* [6].

As a consequence of this electromagnetic induction phenomenon, the magnitude of the current density drops exponentially with the distance away from the surface. The distance at which the current density becomes a fraction $1/e$ of its value at the surface is called *skin depth*, which is calculated by

$$\sigma_s = \sqrt{\frac{\rho}{\pi \mu f}} \tag{1-6}$$

In Equation (1-6), f is the frequency, and μ and ρ are the permeability and resistivity of the material. Making the thickness of the conductor larger than approximately $2\sigma_s$ will not reduce the effective resistance of the line.

Figure 1-16(c) shows the resistance plots over the frequency for the five line configurations shown in Figure 1-16(a). The skin effects are observed at the higher frequencies with the increased resistances for all configurations. Case 4, shown in Figure 1-16(c), which has the largest width, shows the skin effect at the lowest frequency due to its large width.

The impedance of a power line is calculated as follows:

$$|Z(f)| = \sqrt{R^2 + (2\pi f L)^2} \qquad (1\text{-}7)$$

In Equation (1-7), f is the clock frequency and R and L are unit-length line resistance and unit-length line inductance. Figure 1-16(d) shows the impedance as the frequency functions of the V_{cc} and V_{ss} line configurations shown in Figure 1-16(a).

At the high frequency, the impedance is rising, especially for Case 5, due to the inductance effect, as shown in Figure 1-16(d). Case 4, shown in Figure 1-16(a) with the largest wire width and small line space, has the smallest impedance.

The *inductance delay* due to the line inductance and line resistance is calculated as follows:

$$\tau = L/R \qquad (1\text{-}8)$$

The L/R delay characterizes the importance of the inductance in power network modeling. Figure 1-16(e) shows the L/R delay results; Case 2 and Case 3, with small line widths and small line spaces, have the smallest L/R delay, as small as 15–19 ps for a 10000 μm long power line.

If the L/R delay is much smaller than the RC delay per unit length, the line inductance L_{vcc} or L_{vss} can be ignored in the on-chip power network model. In this condition, the RC network is accurate enough to model the on-chip power network.

Based on the experimental results shown in Figure 1-16(e), we can conclude that narrow and dense lines are preferred in the power network design for metal inductance reduction. However other effects, like the IR drop, need to be considered as well.

Just considering how to reduce the inductance effect through wire sizing is not very useful since the inductance is still dominated by the package in modern chips. But we can use dense and narrow lines for reducing both on-chip inductance and resistance. An example is shown in Figure 1-17. The inductance is obviously reduced based on our experiments.

The resistance of these narrow lines combined is equal to, or less than, a wide line. The example in Figure 1-17 shows a practical guideline used in the Intel microprocessor power network design.

1.6 PROCESS SCALING IMPACTS

We have considered two scenarios for the technology scaling in microprocessor chips. Scenario A scales the existing chip to a new process with a scaling factor S with little logic change. In Scenario A, die size is reduced by S^2. Scenario B scales the existing chip to a new process with lots of new logics implemented.

In Scenario B, the die size is assumed to be unchanged when using the new process due to more transistors employed in the new design. Table 1-3 shows the impact on the microprocessor power distribution of using the above two scaling scenarios for the microprocessor chips. The detailed derivations are given below.

Scenario A

The line width and space are both reduced by S, assuming the line thickness change is negligible in process shrinking. The unit-length resistance is increased by $1/S$. The unit-length capacitance is reduced in S by assuming that the plate capacitance is reduced by $1/S^2$ but the coupling capacitance increases by $1/S$ due to the smaller line space.

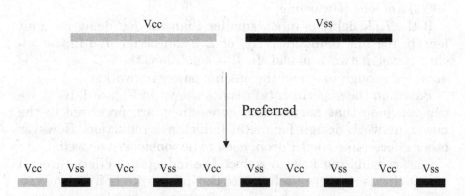

Figure 1-17. Design guidelines for on-chip power lines.

Table 1-3. Technology scaling model for microprocessor power distribution

	Design Parameters	Scenario A	Scenario B
Dimensions	Die size	S^2 (down)	Unchanged
	Transistor count	Unchanged	$1/S$ (up)
	Metal width	S (down)	S (down)
	Metal space	S (down)	S (down)
	Metal thickness	Unchanged	Unchanged
	Global metal length	S (down)	Unchanged
	Decoupling capacitance bound	S^2 (down)	Unchanged
	Area % of decoupling capacitor	Unchanged	Unchanged
RLC Parameters	Metal resistance	Unchanged	$1/S$ (up)
	Metal capacitance	S^2 (down)	S (down)
	Loop inductance	S (down)	Unchanged
	Clock frequency	$1/S^2$ (up)	$1/S^2$ (up)
	Toggling transistors per cycle	Unchanged	$1/S$ (up)
	Average gate capacitance	S^2 (down)	S^2 (down)
	Total gate capacitance	S^2 (down)	S (down)
	Total signal connections	Unchanged	$1/S^2$ (up)
	Total wire capacitance	S^2 (down)	$1/S$ (up)
	Total toggling capacitance	S^2 (down)	Unchanged
Power Consumption	Power consumption (total)	S^2 (down)	Unchanged
	Supply current (total)	S (down)	$1/S$ (up)
	Current density on power line	Unchanged	$1/S^2$ (up)
Voltage Drop	Supply voltage	S (down)	S (down)
	IR drop	S (down)	$1/S^2$ (up)
	$L \cdot Di/Dt$ drop	S^2 (down)	$1/S$ (up)

The die size is reduced by S^2, and the length of power lines is scaled in S. The line resistance for the power network is not changed, and the line capacitance for the power network or long signal lines is reduced by S^2.

Based on Equation (1-5), the unit-length inductance between two adjacent V_{cc} and V_{ss} lines is not changed, because the line space (s) and line width (w) are both reduced by S. The total line inductance is reduced in S, due to the power line length scaled in S.

Chip clock frequency is assumed to increase by $1/S^2$, which is a simplification of the fact that the microprocessor frequency will roughly double every two years for the next process generation. In Scenario A, the logic of the chip is changed very little and the number of toggling transistors per clock cycle is kept unchanged.

The channel length and width of each device are both scaled down in S. The average gate capacitance is down by S^2. So the total gate capacitance is down by S^2. Since the total wire capacitance of signals is also down in S^2, with unchanged transistor numbers and signal connections, the total toggling capacitance ($C_{\text{toggle}} = C_{\text{gate}} + C_{\text{wire}}$) of the chip is reduced by S^2. The supply voltage is scaled in S at each process generation, as shown in Figure 1-1(a).

The power consumption can be estimated as: $0.5 \cdot f \cdot V^2_{dd} \cdot C_{\text{toggle}}$, where f = clock frequency, V_{dd} = supply voltage, and C_{toggle} = total toggling capacitance of the chip. The power consumption is reduced by S^2 based on the above assumptions for the frequency, supply voltage, and the total toggling capacitance per clock cycle.

The current of the power distribution network is calculated by the power consumption divided by the supply voltage. Since the power is down by S^2 and V_{dd} is down by S, the current is thus down by S. Since the line width is down by S and current down by S, the current density of the power line is not changed.

The IR drop is down by S, since the line resistance is not changed but the current is reduced in S. The $L \cdot di/dt$ voltage drop is reduced by S^2 because the line inductance L is scaled down by S; di (current) is reduced by S for the same dt period.

Based on Equation (1-3), we got the bound of the total on-chip decoupling capacitance with 10 times the total toggling capacitance to achieve 10% V_{dd} noise bound. Because the total toggling capacitance is reduced by S^2, the upper bound of the total decoupling capacitance needed in the chip is also reduced by S^2.

Since the die size is reduced by S^2 in Scenario A, the percentage of die size used for the on-chip decoupling capacitance is not changed in this scenario.

Scenario B

The die size is assumed to be not changed in this scenario, so the global line length is not changed. The line resistance of the power network is increased by $1/S$. The line capacitance of the power network, or long signals, is reduced in S, since the unit-length capacitance is down in S, as derived in Scenario A.

Based on Equation (1-5), the unit-length inductance between two adjacent V_{cc} and V_{ss} lines is not changed due to the line space (s) and the line width (w), both reduced by S. The total line inductance is not changed because the global line length is not changed.

The chip clock frequency is supposed to increase by $1/S^2$ about every two years for each process generation. In Scenario B, new logic features are implemented, assuming employment of $1/S$ more transistors in the design. Therefore, the total toggling transistors per cycle increases by $1/S$.

The gate channel length and channel width are both scaled down by S, so each gate capacitance is down by S^2 and the total gate capacitance is down by S. The total signal number is increased by $1/S^2$, for $1/S$ more transistors used in the design. This implies that the total wire capacitance of signals in this chip is increased by $1/S$, based on the unit line capacitance in this scenario being reduced by S.

If we assume that the total wire capacitance is almost equal to the total gate capacitance across a chip (and that is the case we found in a microprocessor chip), we get the unchanged total toggling capacitance, C_{toggle} ($C_{\text{toggle}} = C_{\text{gate}} + C_{\text{wire}}$). The supply voltage is reduced in S at each process generation.

The average power consumption is calculated by $0.5 \cdot f \cdot V_{dd}^2 \cdot C_{\text{toggle}}$, where f = clock frequency, V_{dd} = supply voltage, and C_{toggle} = toggling capacitance. The power consumption is unchanged in this scenario. The current through the power distribution network is calculated by the power consumption divided by the supply voltage. Since the power is unchanged and V_{dd} is down by S, the total current increases by $1/S$.

Because the wire width is down by S and current increases by S, the current density of the power network increases by $1/S^2$. The IR drop increases by $1/S^2$, due to the line resistance increasing by $1/S$ and the supply current also increases by $1/S$. The $L \cdot di/dt$ noise increases by $1/S$ since L not changed; di (current) increases by $1/S$ for the same dt period.

Because the total toggling capacitance per cycle is unchanged, the upper bound of the total on-chip decoupling capacitance is also unchanged, based on Equation (1-3). Since the die size is not changed in Scenario B, the area percentage used for the on-chip decoupling capacitance is also unchanged.

Although the scaling models show unchanged power consumption in Scenario B, for most new microprocessors we see more aggressive transistor number increase or more parallelism used for higher performance. This observation results in more power consumption in new microprocessors. For example, Alpha 21264 (0.35 μm) has 1.63 times more transistors than Alpha 21164 (0.50 μm) (> 1/0.7 = 1.42 scaling factor assumed in Scenario B), and the power consumption is increased from 50 W to 72 W [1].

Process scaling factor S in Table 1-3 is the ratio of the minimum feature sizes between two process generations. S is about 0.7 [10]. For example, an 0.18 μm process is scaled to 0.13 μm for a scaling factor S of about 0.72 (0.13/0.18 = 0.72).

1.7 SUMMARY

This chapter discusses the modeling issues of on-chip power grids. It provides the primary models and characterization results for the resistance, capacitance, and inductance associated with metal lines and vias to route the power distribution network on the chip. The power distribution network, in general, can be characterized as a low-pass RLC filter for the frequency domain analysis.

In addition, the resonant frequency should be removed from the working frequency of the circuit; otherwise, this RLC network will generate a lot of noise. We describe the inductance effects for the on-chip power grid. Usually, very dense and narrow width V_{ss} and V_{cc} lines are interleaved with each other to reduce the inductance.

In general, as a designer of a power grid, you want to increase the capacitance while reducing the resistance and inductance. The latter two parameters are associated with the IR drop and $L \cdot di/dt$ noise.

The capacitance increase for a power grid is implemented by adding intentional decoupling capacitors. In addition, decoupling capacitors are inserted at the noisy nodes of the power distribution network. A CAD algorithm has been proposed to automate this decoupling capacitor insertion process [83].

Finally, we predict future design directions by providing technology scaling models related to power distribution performance and voltage drop based on two different chip improvement scenarios.

2

DESIGN PERSPECTIVES

In this chapter, we describe guidelines for chip layout and floor planning in power grid design. Enough metal power lines should be allocated for both the global power network and local power network in all metal layers in order to deliver current efficiently through the power network. However, power grids or metal lines used for V_{dd} and V_{ss} networks will use up a lot of signal routing resources.

Therefore, there is an intention from the circuit design perspective to ignore the power network metal density at the planning stage in order to reduce the metal layers or reduce the chip size for manufacturing cost reduction, but it carries the risk of increasing IR drop and $L \cdot di/dt$ noise associated with the power distribution network.

Therefore, we believe that planning or design guidelines for the power networks' metal lines are essential at the early design planning stage in order to deliver a successful chip.

This chapter is organized into six sections as follows. Section 2.1 covers power grid planning for a communication chip [45]. Section 2.2 examines power grid planning for two microprocessor chips [46, 47, 48]. Section 2.3 describes the power grid analysis and decoupling capacitance optimization method for another microprocessor chip [49]. Section 2.4 discusses the general methodology for IR drop analysis and reduction. Section 2.5 discusses the package-level power network planning [61]. Section 2.6 is a summary of the chapter.

Power Distribution Network Design for VLSI, by Qing K. Zhu
ISBN 0-471-65720-4 © 2004 John Wiley & Sons, Inc.

2.1   PLANNING FOR COMMUNICATION CHIPS

Deciding on the metal line layout in a chip to minimize the *IR* drop and reduce $L \cdot di/dt$ noise is part of power network planning. Based on Equation (1-1), the metal line resistance is inversely proportional to the metal line width. Based on Equation (1-5), the inductance is also inversely proportional to the metal line width. In addition, based on the guidelines shown in Figure 1-17, the interleaving of V_{dd} and V_{ss} lines in small widths is preferred to reduce the area of the current loop paths and reduce the inductance. In addition, the resistance and inductance are both reduced if we use short metal lines from the power supply pads to the devices.

The methods to improve the layout or package for the power distribution network are summarized as follows:

1. Adding multiple power lines (V_{dd}/V_{ss}) over the chip, usually at some constant space over the chip surface.
2. Adding enough power lines in each layer (for example, M1, M2, M3, M4, M5, and M6, etc.).
3. Adding enough vias between power lines in adjacent metal layers.
4. Using advanced package technology, such as the C4 package, to place multiple C4 power bumps over the chip and to reduce the distance from the bumps to the on-chip power network.

The following design example is from a communication chip, as shown in Figure 2-1 [45].

- The first step is to decide on the floorplanning and chip area. The floorplanning also includes the package options and I/O locations.
- A simplified RLC model is constructed that reflects the power line electrical models. In order to reduce the computational time, the R and C in the area are lumped in the RC model. To improve accuracy, the package model is also included for V_{dd}/V_{ss} pads.
- The inductance may not be included in the above model if it is not significant in the power distribution and the R/L delay is much less than the RC delay, as discussed in Section 1.5.

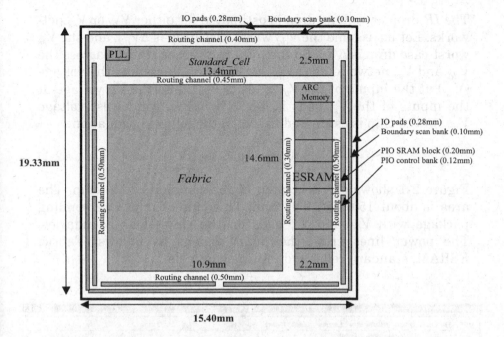

Figure 2-1. Floor plan of a communication chip [45].

- A sensitivity study is executed by varying the metal density or metal widths in the chip floorplanning for the power distribution network. The R and C values in the simplified RC model will be varied based on the density of the power grid metals.

- The sensitivity study can be done by changing the number and locations of the V_{dd}/V_{ss} pads supplied to the chip. We then decide the best IR drop and $L \cdot di/dt$ drop across the chip.

- Once we select the power grid structure, we need to determine the number of V_{dd}/V_{ss} pads and locations and the metal line widths for each layer of power grid.

- Again, the design is optimized for the power grid with regard to the IR drop and $L \cdot di/dt$ drop targets, with as little as possible taken from the layout area.

- The IR drop analysis is performed on the DC analysis for this simplified RC model of the power grid. The package resistance or inductance for each V_{dd} pad is included in the model to analyze the voltage drop across the package.

- The above power grid modeling and analysis should be done for both V_{dd} and V_{ss} networks.

The *IR* drop or voltage drop is estimated for either V_{dd} or V_{ss} networks. Let us assume the V_{dd} worst-case drop is ΔV_{dd}, and the V_{ss} worst-case drop is ΔV_{ss}. So the total worst-case *IR* drop across the V_{dd} and V_{ss} networks is $(\Delta V_{dd} + \Delta V_{ss})$. Let us assume the voltage (V_{dd}) at the inputs of the V_{dd} pads is V_{max}, and the V_{ss} voltage at the inputs of the V_{ss} pads is 0 V. Therefore, the lowest voltage V_{min} in the chip is estimated based on the following equation:

$$V_{min} = V_{max} - (\Delta V_{dd} + \Delta V_{ss}) \tag{2-1}$$

Figure 2-1 shows the floor plan of the communication chip. The area is about 15.40×19.33 mm. This chip is in a wire bonding package with V_{dd} and V_{ss} pads on the chip's four boundaries. The power lines cross the main regions as follows: Fabric, ESRAM, standard cells, and routing channels.

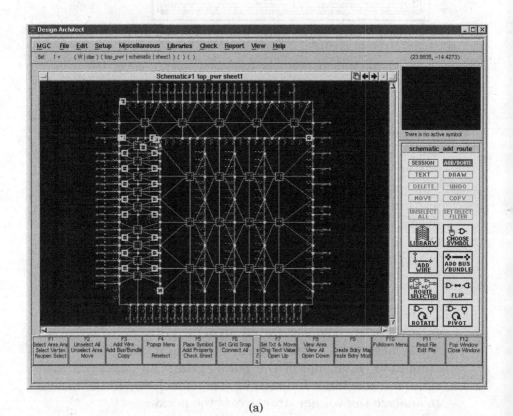

(a)

Figure 2-2. RC modeling of full-chip power grid.

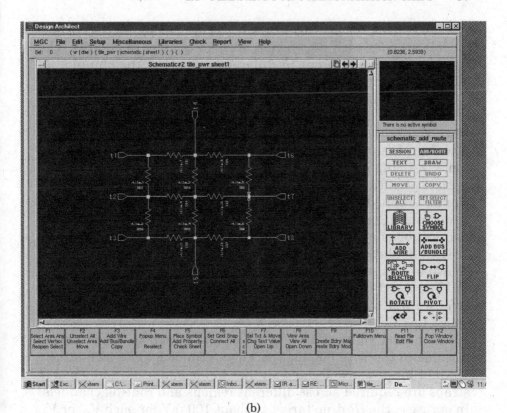

(b)

Figure 2-2 *(continued).*

Figure 2-2(a) shows the simplified *RC* model for the full-chip power grid, and Figure 2-2(b) shows the unit-cell *RC* model for the power grid in each unit region. The entire chip is partitioned into many finer unit regions to cover the power grid. Each node in the unit-cell *RC* model is tied to a current source, which is a DC current to model the average current consumption by the devices located in that region.

The most difficult job in the modeling is to estimate the current consumption, since the current consumption depends on the applications of the chip and it is very hard to determine with accuracy in the model before the chip is manufactured.

There are CAD tools on the market to estimate the current consumption based on test vectors or worst-case assumptions. For a small unit region, we could apply the circuit simulation on the devices to extract the average current. Figure 2-3 shows the current models used for each unit region in this example. In addition, the currents will be different in different regions of the chip due to different circuit density and switching activity.

The modeling of the current sources can be improved continually during the chip design stages as more circuits are designed and more accurate current estimations are obtained. In addition, the power grid current modeling can be further optimized based on some test chip or earlier version chip's power measurement. The initial specifications of the power grid will come up based on the simulation model, as shown in Figure 2-2. Figure 2-4 shows the power routing specifications in the fabric tile region of this communication chip [45].

The simulation result for the power grid model in this chip is shown in Figure 2-5. The simulation is done for the IR drop analysis. The worst-case IR drop, based on Figure 2-5, is about 99 mV (1.71 – 1.6112 V). The lowest ($V_{dd} - V_{ss}$) voltage across the chip is about 1.512 V (1.6112 – 0.0998 V).

For the communication chip power grid design shown in Figure 2-1, due to the wire bonding package technology in which all the V_{dd} and V_{ss} pads are located on the chip boundaries, many power straps are required across different regions and routing channels. In our case, the IR drop target is about 100 mV for each V_{dd} or V_{ss} network across the chip.

The following specifications are given for the power routing on the chip for the V_{dd} network; the V_{ss} network has the same specifications and equal metal lines in the routing [45].

.SUBCKT tile_pwr	.SUBCKT std_pwr	.SUBCKT esram_pwr
I_T1 T1 0 20.3mA	I_T1 T1 0 6.9mA	I_T1 T1 0 0.4mA
I_T2 T2 0 40.6mA	I_T2 T2 0 13.8mA	I_T2 T2 0 0.8mA
I_T3 T3 0 20.3mA	I_T3 T3 0 6.9mA	I_T3 T3 0 0.4mA
I_T4 T4 0 40.6mA	I_T4 T4 0 13.8mA	I_T4 T4 0 0.8mA
I_T5 T5 0 40.6mA	I_T5 T5 0 13.8mA	I_T5 T5 0 0.8mA
I_T6 T6 0 20.3mA	I_T6 T6 0 6.9mA	I_T6 T6 0 0.4mA
I_T7 T7 0 40.6mA	I_T7 T7 0 13.8mA	I_T7 T7 0 0.8mA
I_T8 T8 0 20.3mA	I_T8 T8 0 6.9mA	I_T8 T8 0 0.4mA
I_N_5 N_5 0 81.2mA	I_N_5 N_5 0 27.6mA	I_N_5 N_5 0 1.6mA
.ENDS $ tile_pwr $	.ENDS $ std_pwr $	.ENDS $ esram_pwr $

Figure 2-3. Current consumption in unit regions.

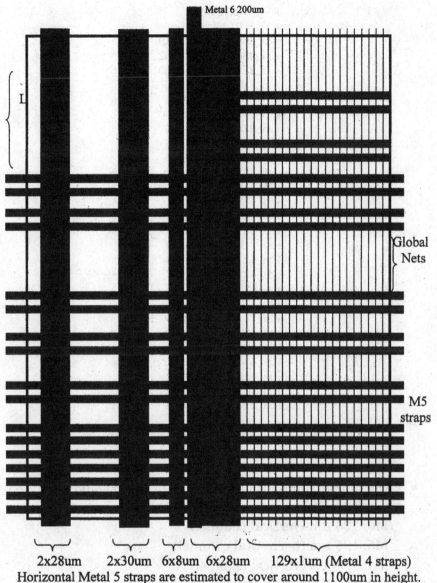

2x28um   2x30um   6x8um   6x28um   129x1um (Metal 4 straps)
Horizontal Metal 5 straps are estimated to cover around 1100um in height.
Metal6 straps of roughly 200um in width will be added in the middle

Figure 2-4. Fabric tile power routing specifications [45].

$$+ V_{dd} = 1.7100$$
$$+ V_{ss} = 0.$$
$$+ xi\_2865.n\_5 = 1.6910$$
$$+ xi\_2866.n\_5 = 1.6932$$
$$+ xi\_2868.n\_5 = 1.6980$$
$$+ xi\_218.n\_5 = 1.6926$$
$$+ xi\_219.n\_5 = 1.6951$$
$$+ xi\_2867.n\_5 = 1.6890$$
$$+ xi\_636.n\_5 = 1.6945$$
$$+ xi\_427.n\_5 = 1.6386$$
$$+ xi\_638.n\_5 = 1.6846$$
$$+ xi\_637.n\_5 = 1.6817$$
$$+ xi\_4.n\_5 = 1.6659$$
$$+ xi\_432.n\_5 = 1.6281$$
$$+ xi\_431.n\_5 = 1.6504$$
$$+ xi\_2870.n\_5 = 1.7029$$
$$+ xi\_840.n\_5 = 1.6959$$
$$+ xi\_424.n\_5 = 1.6503$$
$$+ xi\_423.n\_5 = 1.7007$$
$$+ xi\_428.n\_5 = 1.6112$$
$$+ xi\_425.n\_5 = 1.6508$$
$$+ xi\_434.n\_5 = 1.6529$$
$$+ xi\_430.n\_5 = 1.6439$$
$$+ xi\_429.n\_5 = 1.6120$$
$$+ xi\_433.n\_5 = 1.6261$$
$$+ xi\_426.n\_5 = 1.6692$$
$$+ xi\_6339.n\_5 = 1.6979$$
$$+ xi\_220.n\_5 = 1.7008$$

Figure 2-5. Simulation results of node voltages [45].

- Vertical and horizontal channels between standard cell, fabric, and ESRAM regions (metal width):

    M6: 125 μm (vertical channel)

    M5: 125 μm (horizontal channel)

    M4: 125 μm (vertical channel)

    M3: 125 μm (horizontal channel)

- I/O vertical and horizontal channels between core and pads (metal width):

    M6: 125 μm (vertical channel)

    M5: 125 μm (horizontal channel)

    M4: 125 μm (vertical channel)

    M3: 125 μm (horizontal channel)

- V_{dd} pad connection to core power ring (metal width and length):

    Length: 200 μm

    M6: 90 μm

    M2: 90 μm

    Package resistance for each V_{dd} pad: 40 mΩ (from ball to package substrate to pad)

    Input V_{dd} (lowest) to package V_{dd} ball: 1.71 V

- Fabric tile V_{dd} lines (metal width):

    M6: 200 μm total (vertical) inside the tile, 150 μm total (vertical) added between tiles

    M5: 550 μm total (horizontal) inside the tile, 150 μm total (horizontal) added at sides of tiles

    M4: 230.5 μm total (vertical) inside the tile, 150 μm total (vertical) added between tiles

    M3: 150 μm total (vertical) between tiles, 150 μm total (horizontal) added on two sides of the tile

    M2: 150 μm total (vertical) between tiles.

    M1: 150 μm total (vertical) between tiles, 150 μm total (horizontal) added on two sides of the tile

- Standard cell region V_{dd} lines (metal width):

    M6 completely used for V_{dd} and V_{ss} vertical straps (total M6: ~6.7 mm V_{dd}, ~6.7 mm V_{ss})

    M3: 20 μm width straps (horizontal) per 500 μm space

    M2: 20 μm width straps (vertical) per 500 μm space

    M1: inside standard cells (horizontal) about total 330 μm in the region

- ESRAM region V_{dd} lines (V_{dd} metal width to fill in white spaces):

    M5 completely over the 9 SRAM blocks (ESRAM/ARC) (total M5: ~7.1 mm V_{dd}, ~7.1 mm V_{ss}). 0 μm in channels between ESRAM blocks

    M4: 30 μm ring (vertical) inside each SRAM block

    M3: 30 μm ring (horizontal) inside each SRAM block

Figure 2-6 shows the complete power grid (V_{dd}) simulation model. The node voltages in the simulation by DC analysis are shown in this figure and the lowest voltage is about 1.32 V at the center of

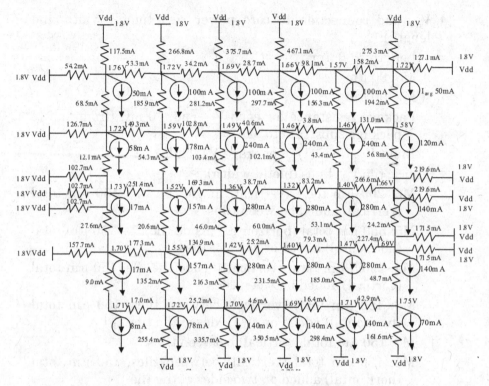

Figure 2-6. Power grid simulation model [45].

the chip. The simplified power distribution model allows us to do the sensitivity study while changing the metal widths and densities in the power routing to see the impacts on the node voltages. The resistance and capacitance of the metal lines are varied, based on the given power routing widths of the V_{dd} network. For example, Figure 2-7(a) shows the lowest voltage at the center of the chip by selecting various metal widths for each V_{dd} or V_{ss} bus in the routing (horizontal × coordinates in the figure) and various metal widths extended directly from each V_{dd} or V_{ss} pad (vertical y coordinates in the figure).

It is done by using parallel metal buses overlapped in the M6 and M4 (vertical) or M5 and M3 (horizontal) layers. Figure 2-7(b) further shows the lowest voltage improvement obtained by adding more parallel buses in M6, M4, and M2 (vertical) and M5, M3, and M1 (horizontal) routing layers. By adding more power buses in M2, compared with Figure 2-7(a) and Figure 2-7(b), the lowest node voltages are slightly improved across the chip by our simulation.

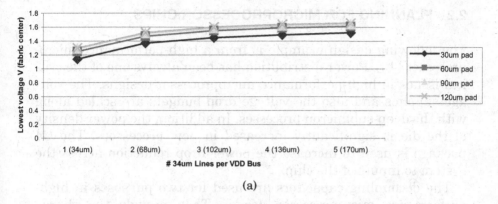

(a)

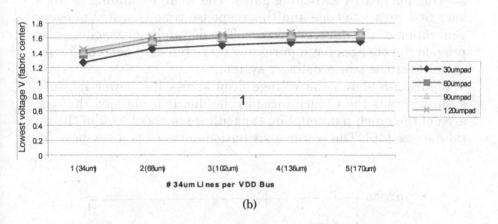

(b)

Figure 2-7. Sensitivity study of power metal widths [45].

The sensitivity study based on the simplified RC model for the entire chip power grid provides a useful tool during the power grid planning. Further sensitivity studies can be iterated during the power grid planning stage to answer the following questions: (1) How many V_{dd} and V_{ss} pads should there be? (2) Where should these V_{dd} and V_{ss} pads be located? (3) Do we distribute them evenly or nonevenly? (4) Do we use wire bonding technology or some other more advanced technology to reduce the IR drop?

In the example we have shown, a huge amount of layout area has obviously been used by the power grid and the chip area will be impacted significantly. So C4 or flip-chip technology is definitely a good alternative for this design.

2.2 PLANNING FOR MICROPROCESSOR CHIPS

The following design example is from a high-performance micro-processor [46]. Power distribution has been always one of the critical issues in high-performance microprocessor designs. The voltage supplies and also the voltage drop budgets are scaled along with the deep-submicron processes. In addition, the power density of the die is significantly increased in new processors. The C4 package is used to increase the power drop reduction across the system to inputs of the chip.

The decoupling capacitors are used for two purposes in high-performance microprocessor design. They provide the charge sharing for nearby switching gates. The local decoupling needs a very fast response time and this response time is scaled in every generation of the microprocessors. The decoupling capacitors also provide the charges for suppressing large full-chip current fluctuations over the power delivery system.

Figure 2-8 shows the voltage drop across the power network system versus the capacitances in the die. It is claimed that the area of the on-chip decoupling capacitance is about 12% of the total die size [46]. The power distribution network is a low-pass fil-

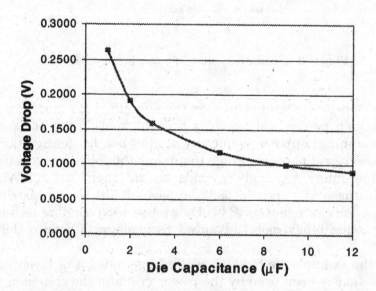

Figure 2-8. Power voltage drop versus decoupling capacitance in a high-performance microprocessor [46].

ter in order to suppress the high-order noise, preferably only for the DC voltage across this system.

Based on the series RLC model, as shown in Figure 1-3, the quality factor Q will be reduced with large C, small R, and small L. Low Q will result in the wide bandwidth needed to allow the AC resonance to pass over the power delivery system. The quality factor of a series RLC network can be determined as follows:

$$Q = \frac{\sqrt{L/C}}{R} \qquad (2\text{-}2)$$

There are two methods to plan the power grid in high-performance microprocessors [46]. The first method uses spreadsheet calculations. It computes the voltage drop for a section of the power grid, which includes the estimation of voltage drops from the package to the transistors.

The second method is to build the complete RLC model of the full-chip and package-level power distribution networks. The full-system models (die, package, and power supply) are needed in the accurate model to perform the voltage simulations across the power network.

It is usually simulated overnight and the model complexity is limited by the simulation time. The results can be used to set the specifications for the power distribution design on the chip and on the package. Here are the detailed steps for the C4-package-based power grid design in the high-performance microprocessor design [46]:

- Start with basic calculations of the current needed for the chip. The current can be scaled from the prior products. It can also be decided on based on the spreadsheet and hand calculations based on the simulation data in individual modules possibly used in the chip.
- Keep in mind that when we design the power grid, the circuit and layout design of each module may not be clear or finalized. So in this stage, a ballpark figure or estimation is used for the power design. Usually, overallocation of the power grid lines are common practice due to the overestimation of the switching current.
- Build the first full system model based on the understanding of what are the causes of the large voltage drop.

- Propose the first-order solution for the die, package, and power supply.
- Develop the initial voltage drop budget and simulation voltage for timing modeling.
- Move toward the detailed design. Determine the exact C4 bump array. Fine-tune the metal grids over the chip based on the detailed *RLC* model's simulation.
- The power grid model is improved during the project when more modules are finalized with circuits and layouts.
- Determine the distance limits of the decoupling capacitors, based on the response time simulation to neighboring switching gates, and eventually come up with the decoupling capacitance placements and sizes needed in the design.

The current estimation usually uses the spreadsheet method, based on power estimates, which have substantial uncertainty [46]. It takes the module power and area into the spreadsheet and produces the map of the power per grid area. The grid area is fit to the C4 bump service area. It converts the power of the current and produces the distribution of the current per bump.

Figure 2-9 shows a detailed M6 grid alignment specification in this high-performance microprocessor. This gives a regular relationship between the two layers. Only two M5 tracks are needed in this assignment to connect M6 to M4 layers. The M6 grid is designed to align with the global M4 grid to enable the efficient routing of the top-level nets and allow for DRC cleaning in the full-chip assembly.

To accomplish this, the following M6 specifications are given for the V_{cc}/V_{ss} lines:

- The M6 grid pitch is a multiple of the M4 grid pitch and will be aligned to the M4 grid on the floor plan.
- The M6 major grid pitch = 538.56 μm, which is 11 times the M4 grid pitch of 48.96 μm. The M6 minor grid pitch = 48.96 μm, which is equal to the M4 grid pitch.
- Each M6 minor grid will exactly overlay the M4 grid under it. The M6 grid is placed on the floor plan such that the Y offset of both major and minor M6 grid is a multiple of 48.96 μm.

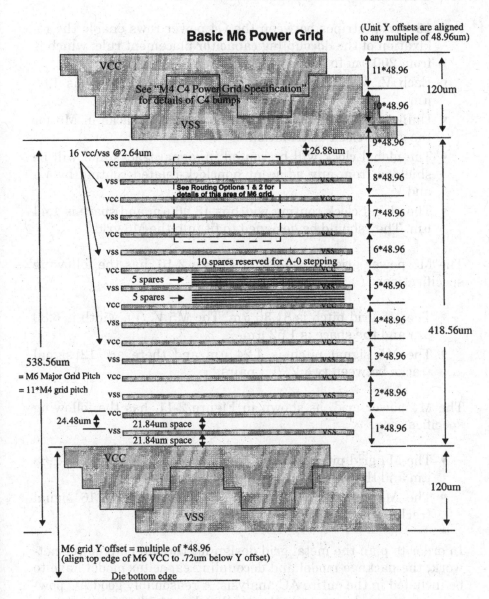

Figure 2-9. Specifications of the power grid on M6 [46].

- 16 V_{cc}/V_{ss} stripes between the C4 power rows enable the relaxation of the decoupling capacitor placement rule, which is from 200 μm to 500 μm.
- Each V_{cc}/V_{ss} strip width is 2.64 μm and the space is 1.68 μm.
- Unlike M4 and M5, there are no reserved tracks in M6 for the global clock distribution.
- The global clock will be routed in signal tracks and will be shielded from any adjacent nonclock-related routing by V_{cc} and V_{ss}.
- The global clock routing width is 18.96 μm and space is 1.44 μm. They should be designed to fit into the M6 grid.

The M5 power grid, as shown in Figure 2-10, has the following specifications:

- The M5 grid pitch is 81.36 μm. The M5 V_{cc}/V_{ss} width is 6.80 μm and the space is 1.52 μm.
- The M5 signal pitch is 4.24 μm and there are 12 signal tracks between two V_{ss}/V_{cc} pairs.

The M4 power grid, as shown in Figure 2-11, has the following specifications:

- The M4 grid pitch is 48.96 μm. The M4 V_{cc}/V_{ss} width is 2.68 μm and the space is 1.04 μm.
- The M4 signal pitch is 2.32 μm and there are 16 signal tracks between two V_{ss}/V_{cc} pairs.

In order to plan the metal grid design for the full-chip power network, the package model and decoupling capacitor model have to be included in the entire AC analysis. A reasonably good AC power network model must be built. We discussed power network modeling and characterization in Chapter 1.

In this section, we will examine the power network AC analysis model from two high-performance microprocessors [47–48]. At the minimum, the analysis must account for the V_{cc} source, the motherboard V_{cc}/V_{ss} traces, the board decoupling capacitors, the CPU socket, the package pin, the power planes, the on-package decoupling capacitances, the CPU I/O, core circuits, and the global clock distribution network.

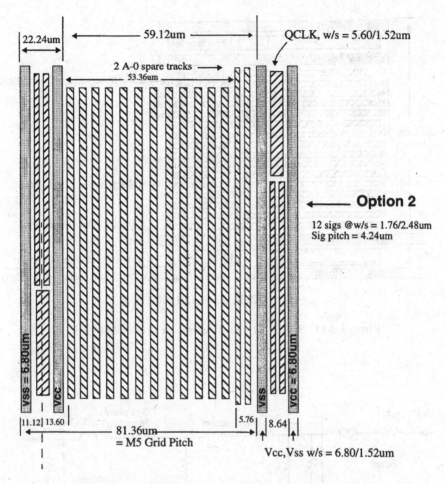

Figure 2-10. Specifications of the power grid on M5 [46].

With this AC model, the CPU I/O and core can be toggled to mimic the execution of the CPU, and the power network performance can be measured and analyzed. The AC model from a high-performance microprocessor is made of three submodels: the package model, the I/O model, and the CPU core model. These models are shown in Figure 2-12.

The I/O and core cell models are represented by an array of the circuit models to model the global power grid on the M4 and M3 layers across the chip, with the switching current tied to each core cell to model the switching activity of the circuit, as shown in Figure 2-13. The current model can be a triangular or other current

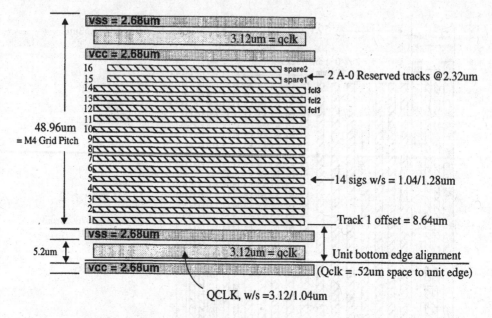

Figure 2-11. Specifications of the power grid on M4 [46].

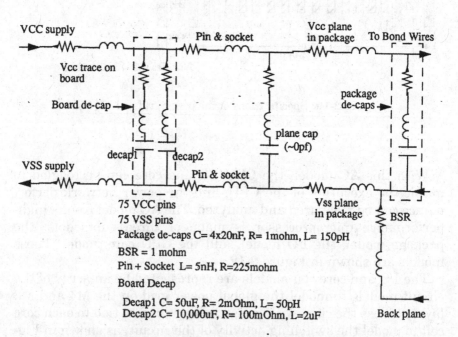

Figure 2-12. Package-level power network modeling [47].

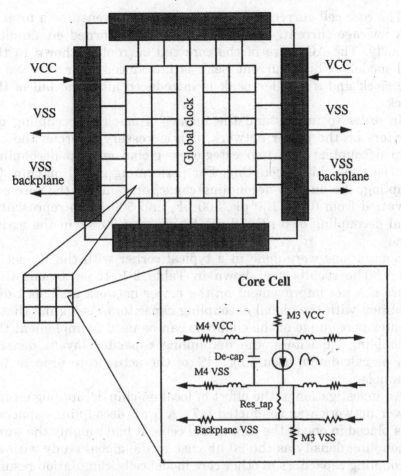

Figure 2-13. I/O and CPU core power network modeling [47].

waveform from the circuit simulation of this design. The I/O model will include the detailed I/O circuits.

Since the global clock tree will consume a lot of power, in this model the detailed model of the clock tree is included for the whole power network simulation. In addition, the decoupling capacitors are included in this model, as shown in Figure 2-13.

As shown in Figure 2-13, the total chip is partitioned into 180 core cells in this AC model. Each cell represents about 1150 × $1000\mu m^2$ of area in the chip. Each cell includes the modeling of M4 V_{cc}/V_{ss}, M3 V_{cc}/V_{ss} and the back power plane network. The on-chip decoupling capacitors are added in the model to simulate the effectiveness of such capacitors.

The core cell current source is turned on to consume a total of 8 A average current. The I/O models can be turned on simultaneously. The AC wave of the core cell current is shown in the cell model. A high current peak is introduced after the rise of the clock and a smaller peak is introduced after the fall of the clock.

In order to understand the impact of on-chip decoupling capacitors on the power network, it is necessary to break the on-chip decoupling into two categories: global on-chip decoupling and local on-chip decoupling. For performing global on-chip decoupling, the on-chip decoupling capacitor value in the core cell is varied from 0 pF, 100 pF, 300 pF, and 500 pF to represent a total decoupling of 0 nF, 18 nF, 54 nF, and 90 nF in the active core.

Simulations were done in a typical corner with the V_{cc} set to 2.5 V. The results are shown in Table 2-1. It is obvious that there is a net improvement on the power network and clock distribution with the global decoupling capacitors. Assuming that a greater percentage of the channels can be used to implement the decoupling capacitors, the decoupling capacitor layout density can be calculated, assuming 34% of the active core area in the channels.

An investigation of the effect of local on-chip decoupling on the power network was conducted [47]. A 5 nF decoupling capacitor was placed in one of the core model cells. It had roughly the same decoupling density as the 90 nF case in the global study with no decoupling capacitors in other core model cells. Simulation results indicate that the effect of the local decoupling is not limited to the cell where the decoupling capacitors are placed. The surrounding core cells, both in the M4 and M3 directions, all benefit from this large decoupling capacitor. The simulation result of this local decoupling is shown in Table 2-2.

Table 2-1. Global decoupling capacitor results [47]

Total decoupling capacitance (nF)	Worst-cycle average V_{cc}/V_{ss} (V)	Worst cycle minimum V_{cc}/V_{ss} (V)	Circuit speed up (gates)	Worst-case global clock jitter (ps)
0	2.071	2.002	Baseline	96
18	2.089	2.046	1.00%	83
54	2.122	2.087	2.40%	74
90	2.136	2.110	2.75%	59

Table 2-2. Local decoupling capacitor results [47]

Total decoupling capacitance (nF)	Worst-cycle average V_{cc}/V_{ss} (V)	Worst cycle minimum V_{cc}/V_{ss} (V)	Circuit speed up (gates)	Worst case global clock jitter (ps)
0	2.051	1.920	Baseline	96
5	2.071	1.959	0.39%	90

The local decoupling capacitors are extremely useful for high-switching-current circuits. They prevent the dip of the power supply voltage around these areas due to the immediate large current flows. For example, if the decoupling capacitors are placed in the left and right I/O areas, ~8 nF total decoupling capacitance in the I/O regions has been reported [47].

The center clock spine will also have decoupling capacitors (~4–5 nF) [47]. It is strongly recommended to have enough decoupling capacitors close to each clock buffer in the chip. The global decoupling is implemented to prevent the overall dip in the power supply. Therefore, the die, the package, and the board design require additional decoupling capacitors for high-performance microprocessors. For example, a minimum of 25 nF decoupling capacitance is required on the die [48]. However, to improve the performance of the power supply network, 60 nF or more is recommended for this processor.

There are usually dead spaces in the layout that are not being occupied by the devices, which may comprise up to 10% of the total die area. In addition, some percentage of the layout area is occupied by the decoupling capacitors based on the AC analysis for the power network. In [48], >1% device of the area is reserved for decoupling capacitors, and >20% of the total area is the channel area used for decoupling capacitors.

As described in Chapter 1, a decoupling capacitor is an nMOS device with its gate tied to V_{cc} and its source and drain tied to V_{ss}. Each μm^2 of the gate area will provide ~5.5 fF of capacitance [48].

There is a set of standard decoupling cells to assist the layout design of the decoupling capacitors, as shown in Figure 2-14 [48]. These standard cells have the split geometries, with split poly contacts and split diffusion contacts, as shown in Figure 2-15 [48]. These standard cells have sizes of 2×2 μm^2, 4×4 μm^2, and 6×6 μm^2.

Fill in any available space with decoupling capacitors. The diffi-

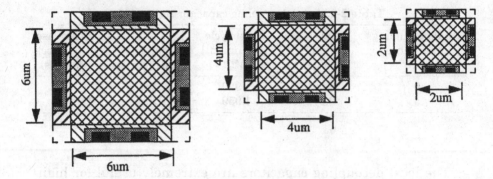

Figure 2-14. Decoupling capacitor standard cells [48].

culty lies in routing the filled decoupling capacitors to the V_{cc} and V_{ss} lines in the layout. Once the decoupling capacitors are inserted into the layout, the schematic should be updated with the inserted decoupling capacitors to make sure the layout versus schematic (LVS) is clean in the layout verification.

When we update the schematic, the decoupling capacitors can

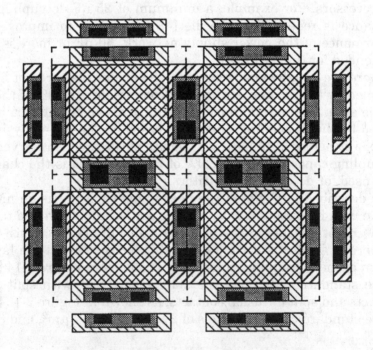

Figure 2-15. Decoupling capacitor layout [48].

add one nMOS device, with the total gate area equal to the sum of all the individual decoupling capacitors.

2.3 IBM CAD METHODOLOGY

A model to analyze the on-chip power supply network of another high-performance microprocessor is described in [49]. A complete power distribution model is shown in Figure 2-16; it includes the package-level power distribution network, the on-chip power bus model, and the equivalent circuits to represent various on-chip switching activities for each functional block.

Among the three major components in the model, the package-level power bus model is dominated by the inductance. The on-chip power bus model is dominated by wire resistances. The switching circuit model determines the switching currents in the chip. In addition, the C_{decap} and R_{decap} in Figure 2-16 show the equivalent model of the decoupling capacitor.

A package-level power bus model for a single-chip site is shown in Figure 2-17. The power and ground distribution networks on the thin-film and ceramic mesh planes are represented with the equivalent inductance model. In this model, the off-chip decoupling capacitors, the multiplayer ceramic vias, the C4 con-

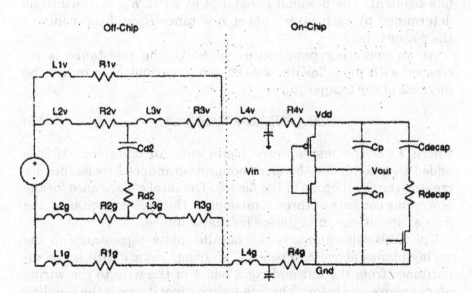

Figure 2-16. Equivalent model for power network AC analysis [49].

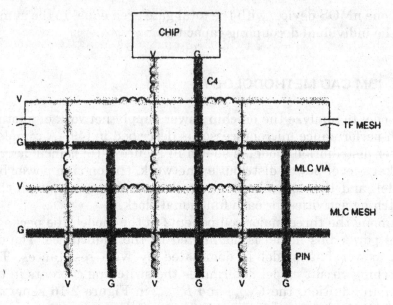

Figure 2-17. Package model of power distribution [49].

nections to the chip, and the I/O pins to the board interface are all included.

To analyze the on-chip power supply voltage drop, we need to model the resistance, capacitance, and inductance of each power bus segment. The nominal resistance at 25°C, $R_{25} = R_s/$width, is determined by each layer's sheet resistance R_s and the width of the power line.

At an operating temperature of 85°C, the resistance is increased with the following well-known linear model to reflect the increase of the temperatures:

$$R_{85} = R_{25}[1 + T_c(85 - 25)] \qquad (2\text{-}3)$$

where T_c is the temperature coefficient. An additional 10% is added to account for the electromigration-induced resistance increase over the lifetime of the device. The total capacitance for the power bus consists of three components: the area capacitance, the fringe capacitance, and line-to-line capacitance.

The area capacitance is the parallel plate capacitance to the wiring planes above and below. The fringe capacitance is the capacitance from the left and right edges of the wire to the wiring planes above and below. The line-to-line capacitance is the coupling capacitance between adjacent wires on the same wiring plane.

The inductance modeling is more complex, since the formula is not well developed. Therefore, an impedance characteristics program is usually used to calculate the inductance [50].

An equivalent RLC power bus network can be generated. In order to reduce the complexity for full-chip analysis, a hierarchical approach is used to build the on-chip power bus model. At the chip level, a global routing grid is generated.

In order to reduce the complexity of full-chip analysis, a hierarchical approach is used to build the on-chip power bus model. At the chip level, a global routing grid is generated to subdivide the chip into global routing cells. All the switching activities within one global routing cell are lumped together, and adjacent cells are connected in global power buses.

At the macro level, where local hot spots are located, a finer grid will be generated to model the detailed power bus structure. Since the power supply voltage in one region can be affected by the switching activities in the neighbouring regions, the finer detailed power bus model should always be connected to the adjacent global power bus model to ensure the analysis results.

It also confirmed that the excessive power supply drop ΔV in the deep-submicron design also necessitates the use of the on-chip decoupling capacitors in addition to the off-chip decoupling capacitors. Without any decoupling capacitors, the impedance will be as follows:

$$Z = R + j\omega L \tag{2-4}$$

where R is the resistance, L is the inductance of the power distribution network, and ω is the angular frequency.

Obviously, the impedance is increased linearly with the frequency in this case, and more ΔV across the power distribution network will be observed in the high-frequency applications.

To model the switching activities for each functional block, we build an equivalent circuit, which consists of time-varying resistors (R_1, R_2, R_3), loading capacitors (C_1, C_2, C_3) and decoupling capacitors (C_{d1}, C_{d2}), as shown in Figure 2-18(a). The loading capacitance for the equivalent circuit is calculated by $C_L = P/(0.5\ V^2_{dd}f)$, where P is the estimated power for the corresponding area, V_{dd} is the power supply voltage, and f is the clock frequency.

When the circuit is switched off, the time-varying resistance will be set to R_{off}. Since not all circuits will switch at the same time, the circuit represented by the loading capacitance C_L can be further partitioned into subcircuits represented by $C_1, C_2, C_3, \ldots$, where the total capacitances will be C_L in order to simulate the distributed switching activities. The timing and delay patterns of each subcircuit can be controlled separately by switching on and off $R_1, R_2, R_3, \ldots$ at different times.

If the simulation results of the functional blocks are available, we can replace the nonlinear devices and capacitive loads in the switching-circuit model with the piecewise linear current sources, which mimic the waveforms of the actual circuits.

A triangular or trapezoidal current waveform, which is simpler than the piecewise linear current waveform, can be derived by calculating the total average current I_{ave} and peak current I_{peak} for each macro in the procedure listed, as follows [49]. The triangular and trapezoidal current waveforms are shown in Figure 2-18(b).

- Simulate the circuit without loading to obtain the internal I_{ave} and I_{peak}.
- Calculate the total output capacitance C_{out} from all output nets.

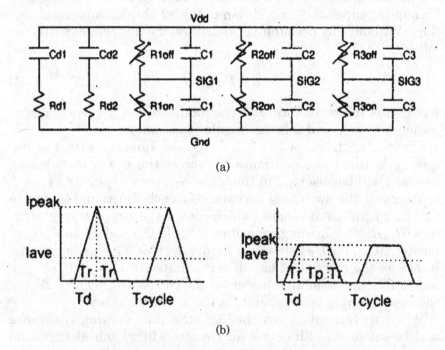

(a)

(b)

Figure 2-18. Switching model for power network simulation [49].

- $I_{ave}(\text{total}) = I_{ave}(\text{internal}) + C_{out} \cdot V_{dd} \cdot f$, where V_{dd} is the power supply voltage and f is the frequency.
- $I_{peak}(\text{total}) = I_{peak}(\text{internal}) \cdot n$, where n is an empirical ratio between the peak current with loading and the peak current without loading.
- Calculate the total power using the following formula: $P = 0.5 \cdot V_{dd} \cdot [I_{ave}(\text{internal}) + C_{out} \cdot V_{dd} \cdot f \cdot SF]$.

After the equivalent circuit of each functional block is generated, it will then be assigned to the global routing cells where the functional block is located, and connected to the corresponding points on the power bus. The model for the on-chip decoupling capacitors consists of three major components: the n-well capacitor C_{nw}, the circuit capacitor C_{ckt}, and the thin-oxide capacitors C_{ox}. The n-well capacitor C_{nw} is the reverse-biased PN junction capacitor between the n-well and p-substrate, as shown in Figure 2-19(a).

The time constant for C_{nw} is process-dependent, but usually can be characterized as between 250 ps and 500 ps. The circuit capac-

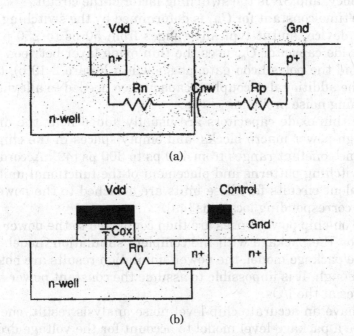

(a)

(b)

Figure 2-19. Decoupling capacitors and RC modeling [49]. (a) n-well junction capacitor. (b) Thin oxide capacitor.

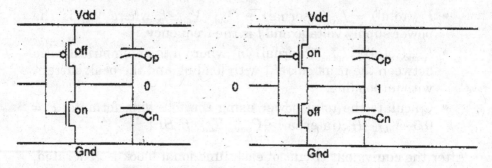

Figure 2-20. Switching capacitor provided by a nonswitching circuit [49].

itor C_{ckt} is derived from the built-in capacitance between V_{dd} and ground in nonswitching circuits, as shown in Figure 2-20. The total capacitance C, the sum of C_p and C_n, from nonswitching circuits, is estimated as [49]:

$$C = [P/(V^2 f)] \cdot (1 - SF)/SF \qquad (2\text{-}5)$$

where P is the power of the circuit, V is the power supply, f is the frequency, and SF is the switching factor of the circuit.

The time constant for C_{ckt} is determined by the switching speed of the device, and it typically ranges from 50 ps to 250 ps. The thin-oxide capacitor C_{ox} uses the thin oxide layer between the n-well and the polysilicon gate, as shown in Figure 2-19(b), to provide the additional decoupling capacitance needed to alleviate the switching noise problem.

The thin-oxide capacitors are usually added near the drivers, the high-power macro blocks, and empty spaces in the chip. The RC time constant ranges from 100 ps to 300 ps [49]. According to the switching patterns and placement of the functional units, the equivalent circuits for these units are attached to the power bus in the corresponding locations [49].

The on-chip power buses are then connected to the power structures on the package with the complete simulation model. Without the package model, the power simulation results are not accurate enough. It is impossible to assume the constant power supply voltages at the I/Os.

To have an accurate chip-level noise analysis result, one must include a package-level model to account for the voltage drops on both the package level and chip level.

Signals can be switched with some patterns for a long time,

with different impacts on the power supply voltage. The difficulty lies in the timing patterns that must be extracted accurately in order to simulate the dynamic switching supply voltage waveforms. In a lot of cases, the power network is overdesigned to accommodate the worst-case switching patterns of the circuits or signals in each functional block.

It is even more important to model the switching patterns between functional blocks correctly. We are concerned with not only the steady-state noise of the hot spots, but also the transient noise when circuits switch from one power level to another. To examine the different noises between units in the chip, the authors in [49] partitioned a chip into nine (3 × 3) regions.

With a power supply voltage of 2.5 V in a 0.25 μm CMOS technology, and when circuits are switched from 20% idle power to 100% full power, the transient voltage and the steady-state voltage are measured in each region. If the flip-chip or C4 technology is used to provide the on-chip power supply, the minimum steady-state V_{dd} will be about 2.37 V [49].

If using wire-bonding peripheral I/Os instead of the C4 technology, the minimum steady-state V_{dd} in the center region will drop to 2.0 V. The following section describes a decoupling capacitor optimization procedure to minimize the sizes and optimise the locations of the on-chip decoupling capacitors with the floor-planning constraints [49].

Most designs now require the voltage drop to be within 10% of V_{dd}. To achieve this goal, decoupling capacitors are added to minimize the switching noises. For high-performance circuits with a frequency of 400 MHz or higher, 10% or more chip area is needed for this purpose. Therefore, it is important to estimate and allocate the area needed for on-chip decoupling capacitors during the early floor-planning stage.

The floorplanning of decoupling capacitors is restricted by the topological and ordering constraints of the preplaced functional blocks. Two directed acyclic graphs are used to represent the vertical and horizontal spaces between adjacent blocks. The edge weights in the acyclic graphs represent the spaces allocated between adjacent functional blocks [49].

The optimization of on-chip decoupling capacitors involves an iteration process between the circuit simulation and floor planning. Given the specifications and locations of each function block, the circuit simulator will analyze the switching noise of the power bus, identify the hot spots, and then determine the amount of decoupling capacitance C_n needed for each global cell n in the power grid.

The floorplanner then translates the amount of decoupling capacitance into physical area A_n, generates pseudoblocks in each region, and determines their locations and dimensions. The added decoupling capacitors will be modeled and simulated in the new floor plan during the next iteration until ΔV is satisfied [49].

2.4 DESIGN FOR *IR* DROP

IR drop is a reduction in voltage that occurs on both power and ground networks in integrated circuits. Narrower metal line widths cause an increase in the metal resistance and therefore in the amount of the voltage drop in the chip. The amount of the voltage drop depends on the effective resistance from the power pads to the logic gates. The metal-line resistance is formulated in Equation (1-1).

The voltage drop is calculated by the following formula:

$$\Delta V = I_{\text{avg}} \cdot R_{\text{eff}} \tag{2-6}$$

where I_{avg} is the average current switched by the logic gates from the power lines originating from a V_{dd} pad. The term *IR* drop (ΔV) is derived from Equation (2-6), which is based on the product of the current I flowing through the effective resistance R_{eff}. Based on Equation (2-6), the methods to reduce the voltage *IR* drop are summarized as follows:

- Reducing the current consumption (I_{avg}) of logic gates. Therefore, any low-power design techniques on the circuit will help. Process scaling or capacitance reduction will also help.
- Another alternative is to increase the number of V_{dd} and V_{ss} pads in the chip to reduce the current consumption for each pair of V_{dd} and V_{ss} pads.
- If the gates along the metal line switch together, the IR drop can be larger due to the increased I_{avg}. Therefore, some alternative switching order for large current gates helps to reduce the IR drop.
- Reducing the wire resistance. In this category, the widening of the metal lines for power lines, or adding more power lines

in the layout are obviously preferred in the power grid floor plan.

- In addition, multiple power layers with extremely dense power lines in the layout are used for high-performance microprocessors. The wire resistance is also proportional to the metal line length from the power pads to the logic gates.
- The C4 package technology provides the area I/O pads, which can provide short power lines. Therefore, most high-performance chips currently use the C4 technology instead of the wire-bonding technology for this reason.

Figure 2-21 shows a power supply connected to the chip pads. The resistors in this figure are the effective resistances in the V_{dd} and V_{ss} power grid distribution. R11–R14 are for V_{dd} and R21–R24 are for V_{ss}. G1–G4 are for logic gates. When the designers are doing the transistor-level simulation, the voltages (V1–V4) are assumed to be equal.

In reality, due to the power grid resistances, the V_{dd} voltage will be reduced due to the current flowing through resistors R11–R14, whereas the V_{ss} voltage will be increased due to the same current flowing through resistors R21–R24. The worst-case drop between the V_{dd} and V_{ss} at any logic gate G1–G4 should be estimated as follows:

$$\Delta V_{\max} = \Delta V_{dd} + \Delta V_{ss} = I_{\text{avg}} \cdot R_{V_{dd}} + I_{\text{avg}} \cdot R_{V_{ss}}$$

or

$$\Delta V_{\max} = I_{\text{avg}} (R_{V_{dd}} + R_{V_{ss}}) \tag{2-7}$$

where $\Delta V_{\max}$ is the worst-case voltage drop between V_{dd} and V_{ss}, ΔV_{dd} is the IR drop of the V_{dd} distribution, and ΔV_{ss} is the IR drop

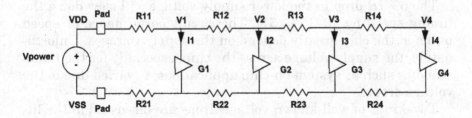

Figure 2-21. Power grid modeling [51].

of the V_{ss} distribution. I_{avg} is the average current consumption of the region provided by one pair of V_{dd} and V_{ss} pads. The ($R_{V_{dd}}$ + $R_{V_{ss}}$) is the sum of effective resistances in the V_{dd} and V_{ss} distribution lines from the pair of V_{dd} and V_{ss} pads to their supplied logic gates.

The IR drop can either have a local or global effect on the chip performance [51]. The *IR* drop is a local phenomenon when a number of gates in close proximity switch at once, causing the *IR* drop in that area. A local *IR* drop can also be caused by a higher resistance to a specific portion of the grid, such as R14 being much larger than expected.

The IR drop can also be a global phenomenon when activity in one region of a chip causes an *IR* drop in other regions. In a well-meshed power grid with equally distributed currents, the power grid typically has a set of equipotential *IR* drop surfaces that form concentric circles cantered in the middle of the chip. So the center of the chip usually has the largest *IR* drop or the lowest supply voltage, especially in the wire-bonding technology. The *IR* drop formula illustrates that it is important to model the switching patterns of the logic gates in a continuous timing period.

If all the gates switched at once, the local or global drop on a chip would be extremely large, an example being when the clock and synchronized elements are switched at the same time. The peak *IR* drop is much larger than the average *IR* drop. The peak *IR* drop happens in the worst-case switch patterns of the logic gates, which excite the maximum amount of power from the gates.

The primary cause of the simultaneously switching *IR* drop is the gate switching due to the clock, the bus, or signal pads. When the global drop is high, but not high enough to cause logic failure, the *IR* drop may cause the timing failure. The *IR* drop, which lowers the supply voltage, will slow down the speed of the gate operation.

The 5% *IR* drop in the lower supply voltage will slow down the timing speed by 10–15% [51]. The circuit performance or speed paths in the chip greatly depend on the supply voltages. Unfortunately, the supply voltage across the chip, especially for the large-size dies such as system-on-chip applications, is varied due to the voltage drops.

Two kinds of well-known voltage drops are discussed in the literature for the on-chip power supply network: *IR* drop and di/dt

noise [6, 52]. The *IR* drop is defined as the average of the peak currents in the power network multiplied by the effective resistance from the power supply pads to the center of the chip. Therefore, in the wire-bonding environment, we can observe the worst-case *IR* drop or the lowest supply voltage at the center of the chip.

Flip-chip technology, which provides area pads on the top of the chip, can ease this problem and this package technology is seen to be more popular for the chips employing 0.13 μm process technology due to the *IR* drop problem.

The following example shows the *IR* drop problem in a wire-bonding package technology with five metal layers with 0.25 μm process technology. M5 is completely used for power straps to reduce the *IR* drop. Readers can see the severity of the *IR* drop problem in the case of the wire-bonding package technology in the communication chip.

A postlayout simulation methodology has been described as follows [54]. The methodology has been used in the standard cell design style in a V_{dd} and ground mesh structure, as shown in Figure 2-22. The standard cell design style has the regular rows of cells aligned in multiple rows, and the power lines of the standard cells are butted together in the same row. The circuit simulation to a set of standard cells is used to understand the parameters that impact the *IR* drop.

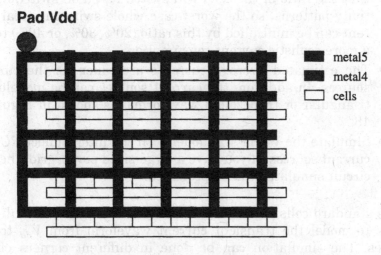

Pad Vdd

metal5
metal4
cells

Figure 2-22. Power mesh on standard cell design [54].

Knowing when and under which conditions the currents to the standard cells are large, we can devise the following method to simulate the most severe *IR* drop.

- Simulate all standard cells and classify them into two classes: negligible *IR* drop impact and severe *IR* drop impact. The latter class for all the standard cells will have current from the V_{dd} to the cell at the switching points greater than the current threshold (i.e., 1 mA).

- Draw the schematic of the V_{dd} mesh, featuring a metal resistor for each vertical or horizontal metal segment of the power mesh. It is recommended that a contact or via resistance be inserted in order to improve accuracy. In the postlayout, the *RC* extraction tool can be used to get the complete *RC* network [59, 60].

- At each cell of the power grid, add a current source to model the sum of the switching current of cells tied from this point.

- Partition the whole chip into smaller areas based on the current source points in the above modeling. Inside each area, we can calculate the average current from V_{dd} to all cells belonging to the area.

- A worst-case assumption can be made that all the cells in this area will switch at the same time if we do not have the switching activity patterns. But the best way is to decide that the ratio of the cells will switch based on switching activity patterns, so the worst-case whole switching total current can be multiplied by this ratio (20%, 30%, or 40%) to get a more realistic current consumption.

- The estimated average currents are taken as the current sources. In addition, the current sources can be modeled as triangular or trapezoidal waveforms, as shown in Figure 2-18.

- Simulate the V_{dd} or V_{ss} model with the interconnect *RC* and current sources. If you have a large-sized power grid, the fast circuit simulator will be preferred.

The standard cells simulation can be done using the stimuli vectors to model the transient current waveform from V_{dd} to the gates. The simulation can be done in different corners of the process, with different temperatures, supply voltages, and transi-

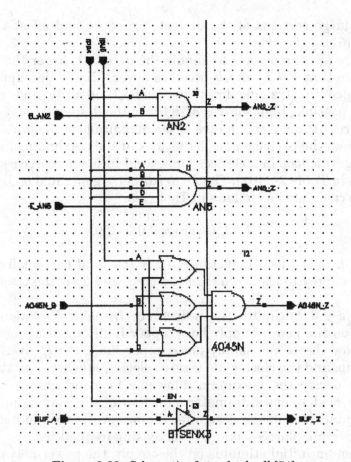

Figure 2-23. Schematic of standard cell [54].

tion times of the input signals to the standard cells. Figure 2-23 shows the schematic of a few standard cells in the design [54].

2.5 PACKAGE-LEVEL METHODOLOGY

There is a general trend toward higher and higher on-chip di/dt noise and less and less tolerance for the voltage noise caused by the fast switching currents ($L \cdot di/dt$). Many factors are making the di/dt problem worse: faster transistors, high current levels, shorter clock cycles, lower noise tolerance due to lower V_{cc} levels, and power saving techniques. Low power design techniques actually degrade the stability of the on-die power supply levels because large sections of the die get turned on and off at various times [61].

There are three ways to handle the di/dt: (1) lower the inductance so that $V = L \cdot di/dt$ becomes lower, (2) add decoupling capacitance in strategic locations, and (3) identify and reduce, where possible, high sources of di/dt in the design.

In order to get a rough idea of the magnitude of the problem, as seen from the package pins, let us look at the maximum allowable package–die loop inductances for several Intel microprocessors, as shown in Table 2-3 [61]. The $L \cdot di/dt$ noise generated on the chip can be calculated as follows in Table 2-3:

$$L \cdot I_{cc}(\text{average})/(0.5 \cdot T_c) \tag{2-8}$$

where L is the loop inductance, I_{cc} is the total current from the power supply to the circuits of the chip, and T_c is the clock cycle time.

Table 2-3 calculates the inductance L, using Equation (2-8), based on the power supply noise upper limit, about 5% of V_{dd}. If we know the power supply noise upper limit, the $I_{cc}(\text{average})$ of the chip, and the clock cycle time or clock frequency, Equation (2-8) can derive the maximum allowable loop inductance L. This simple model shows dramatic reduction of the maximum allowable inductance in the design for the power network in high-performance microprocessors with increasing frequencies.

Given an initial stimulus on the circuit, the power network V_{cc} and V_{ss} will try to oscillate 180 degrees out of phase at the ringing frequency as follows:

$$\omega_0 = \frac{1}{\sqrt{LC}} \tag{2-9}$$

Table 2-3. Maximum allowable inductances to achieve power noise limits in high-performance microprocessors [61]

Frequency (MHz)	di/dt	Power Noise Limit	Maximum Allowable Inductance, L
100	3 A/5.0 ns	165mV	275 pH
150	7 A/3.3 ns	145mV	68 pH
200	7 A/2.5 ns	125mV	45 pH
300	7 A/1.6 ns	90mV	21 pH
500	40 A/1.0 ns	90mV	2 pH

where L is the total power supply loop inductance, and C is the V_{cc}/V_{ss} total capacitance, including the decoupling capacitance inserted in the design. The oscillator may be forced to oscillate at the device's clock frequency if the current levels are high enough. The magnitude of the oscillation is referred as the power supply noise level V_{noise}, as shown in Figure 2-24.

V_{noise} is related to many factors in the design, and is mainly based on the following: (1) the power supply inductances for V_{dd} and V_{ss}, (2) the V_{dd}/V_{ss} on-die capacitance C_{die}, (3) the power supply resistance, and (4) the di/dt numbers from the switching gates [61].

The following are common techniques in microprocessor circuit design to reduce the power noise levels.

- Supply the chip with as many V_{dd} and V_{ss} pins as possible to reduce the $L_{V_{cc}/V_{ss}}$ loop inductance.
- Add the decoupling capacitors on the die so that the highest frequency components of di/dt do not need to be supplied by highly inductive paths through the package and board.
- Try different architecture techniques to limit di/dt, especially in the case of clock gating for power saving.

The minimum and maximum of V_{dd} and V_{ss} have performance and reliability implications, respectively. Timing slowdown may occur when V_{dd}/V_{ss} is at a minimum. Timing skews may arise from some circuits speeding up at high V_{dd}/V_{ss}, and others slowing down at low V_{dd}/V_{ss}. Hot electron operating limits or gate oxide stress limits may be exceeded during the V_{dd}/V_{ss} peaks, leading to reliability failures.

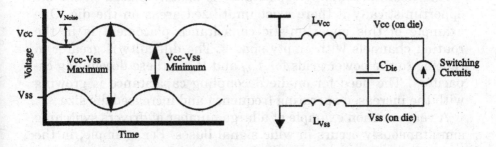

Figure 2-24. LC Oscillation due to power distribution [61].

The timing failures are easy to catch during testing, but reliability problems are not. Low-power design introduces its own set of problems. An ideal low-power design would result in low values of I_{avg} and di/dt. All units on the die would use small currents when active and very little current when inactive.

Low-power designs for microprocessors can typically result in reducing the maximum current peaks moderately, reducing the time spent at peak levels greatly, and causing very low values of current when the chip is carrying out easy tasks or is in standby mode [61].

One concern is the use of lower voltage to achieve low power. Although low power supply voltages help lower the power consumed, higher transistor counts and higher frequency rates usually keep the I_{cc} relatively high.

Lower V_{cc} usually means maintaining a lower absolute value of the voltage noise. Considering the IR drop across the die, power supply guard bands, and tester guard bands, very little margin is left for the on-die power supply oscillations. Since the di/dt usually remains fairly high, large values of decoupling capacitance are needed.

Decoupling capacitance reduces the power supply noise by charging up during the steady state and supplying current during the time at which the circuit switches. Also, decoupling capacitance filters out the differential mode noise on the V_{ss} line from the power supply by keeping the V_{dd} and V_{ss} constant.

Some amount of decoupling capacitance exists naturally on the chip—capacitance of n-wells to the substrate, capacitance of the circuits that are not switching, capacitance between the V_{dd} and V_{ss} traces, etc. A conservative estimate is that only 10–20% of the circuits on the chip switch at any given time; the remaining circuits act as decoupling capacitors [61].

Additional decoupling capacitance is usually placed on the die opportunistically if there exist unutilized areas on the die. One example of this opportunistic capacitance placement is in the routing channels with empty spaces. The difficulty is greater in routing to the power grids for V_{dd} and V_{ss} to these decoupling capacitors. The need for on-die decoupling capacitance is growing with the increased operating frequency and increased die size.

A very common example of a large number of drivers switching simultaneously occurs in wide signal buses. For example, in the case of a microprocessor, the worst-case scenario is with the write-back bus on four different ports, for a total 292 bits switching simultaneously. Each bit drives a 5 pF load with a CMOS inverter size of pMOS = 120 μm and nMOS = 78 μm.

Figure 2-25 shows a plot of the maximum supply voltage drop as a function of the total width of a p-transistor switching simultaneously from low to high for this write-back bus. The write-back bus drivers are laid out in a strip 1000 μm tall and 6000 μm long [61].

The power supply noise is obtained by simulating bus drivers in a power grid model for this microprocessor, with the resistance and inductance of lines and decoupling capacitors properly modeled. In Figure 2-25, the amount of the decoupling (C_D) related to the total load (C_D/C_{load}) is varied to show the effects on the power supply noise [61].

Identifying potential noisy areas on the die based on the locations of wide signal buses is fairly easy. However, it is not an easy task to find clumps of simultaneously switching random logic gates on the die. Such clumps as commonly used can be as bad as the example given above in terms of injecting noise into the supply rails. Hot spots can be identified by summing up the driver sizes (pMOS only or nMOS only) that switch in the same timing window from adjacent devices in the design.

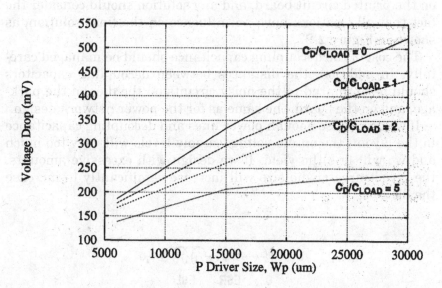

Figure 2-25. Voltage drop versus driver size and decoupling capacitance [61].

From the above discussion, it is apparent that for low supply noise, oversized drivers should be avoided. The driver should be sized just big enough to meet the timing goal. In fact, a slightly undersized driver may be faster than an oversized driver, because of higher supply voltage available to the undersized driver during the switching due to lower supply noise.

A more accurate model for the decoupling capacitor is shown in Figure 2-26. It takes into account the lossy ESR (effective series resistance) and inductive ESL (effective series inductance) properties, as well as the actual capacitance value.

When used to decouple the V_{cc}/V_{ss} voltage planes, this model needs to be modified to add the effective inductance of interconnects (vias) and the plane segment connecting the capacitor to the load. Inductive levels are most significant in high-speed decoupling applications. The lossy component, represented by the ESR of the capacitor, is most significant in decoupling large current transitions such as those around a high-power voltage regulator.

With a lower absolute voltage margin and increasing load dynamics, the ability of the system power supply to directly power the CPU becomes quite limited. To avoid excessive *IR* and inductance-generated voltage drops, a DC/DC converter is used to power the CPU.

Decoupling capacitance is added on the die, in the package, and on the printed circuit board, and any solution should consider the fact that all locations have an influence on the final solution, as shown in Figure 2-27.

The cost of the decoupling capacitance should be managed carefully. In addition, the distances between decoupling capacitors should be optimized to the noisy circuits on the die, on the package, and on the board, the same as for the power network design.

If we do not use enough power lines and decoupling capacitance in the layout, the on-die voltage supply levels will vary too much and we will lose the yield. If we design with excessive amounts, the layout area or die size will increase significantly to increase the die cost.

Figure 2-26. Model of decoupling capacitor [61].

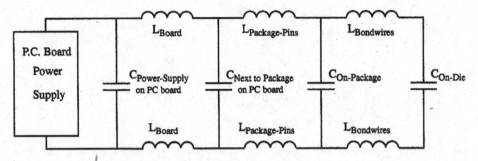

Figure 2-27. Hierarchy of power distribution and decoupling capacitance [61].

2.6 SUMMARY

Power network planning is discussed in this chapter. The power network plan step specifies the metal lines (widths, pitches, etc.) and decoupling capacitor locations for the power distribution network in the chip. The power network is implemented in each metal layer of the die, the package, and the system board.

The design guidelines should be optimized and specified for the metal lines and decoupling capacitors on the die, the package, and the system board. In order to achieve that, the complete *RLC* network is usually constructed for the prelayout metal lines used for the power network. In addition, the decoupling capacitors are included in the modeling, as well as the package models.

High-performance microprocessor design usually employs this kind of optimization study in order to provide accurate specifications of the metal lines for the power distribution. The difficulty in power network modeling is the current waveform modeling to simulate the transistor switching activity in the design.

Usually, simplified triangular or trapezoidal waveforms are used to model the switching currents. The timing patterns of the circuit switching are also important to capture the dynamic (not the worst-case) current consumption in the design.

3

ELECTROMIGRATION

Electromigration in an IC is the movement of metal ions as the result of the flow of electrical charges through the metal wires in the chip, particularly the wires that distribute the power within the IC. This unwanted ion movement could open up metal voids in some parts of the wires and build up metals at other sites.

At the sites from which metal migrates, voids increase the resistance of the affected wire and, in extreme cases, can cause it to open completely. At the receiving end of the migration path, the buildup of metal can form hillocks that, in extreme cases, can span the gap between adjacent wires and cause shorts between them [5].

This chapter is organized into four sections as follows. Section 3.1 discusses the basic definitions and rules for IC electromigration reliability. Section 3.2 describes the CAD tool used to perform the electromigration (EM) analysis [65]. Section 3.3 further discusses the design methodology for reducing IC electromigration failures. Section 3.4 summarizes the chapter.

3.1 BASIC DEFINITIONS AND EM RULES

The increase in resistance caused by electromigration appears only after a period of incubation. During this period, wire resistance remains fairly constant. After that, it increases steadily, eventually causing the IC to fail. How long incubation lasts is determined by such factors as wire size and composition, as well as the current density.

Power Distribution Network Design for VLSI, by Qing K. Zhu
ISBN 0-471-65720-4 © 2004 John Wiley & Sons, Inc.

In process technologies below 0.18 μm, IC metal lines are usually formed of aluminium, some alloy of aluminium and silicon, or aluminium and copper. Pure aluminium has low resistivity, but it is also the most susceptible to electromigration. Copper, which has much lower and better resistance to electromigration, is usually used in 0.18 μm and below processes.

Information obtained during the accelerated testing of IC chips is used for predicting the IC mean time to failure (MTF) under normal operating conditions. The overall relationship of all factors under DC conditions contributing to MTF can be described using Black's equation, as follows [5]:

$$MTF = (AJ^{-N})e^{E_a/kT} \tag{3-1}$$

where J = current density, E_a = activation energy, k = Boltzmann's constant, and A = an experimentally determined scaling factor.

For dynamic operation of a circuit, the equation can be modified by replacing current density, J, with an effective current density, J_{eff}. A factor is adjusted, based on the experimental measurement data, to fit Black's equation curve with the reliability data from the measurements.

Although electromigration is a serious problem in submicron designs, it seldom affects a small portion of the design. In most cases, it is limited to the power distribution network. The problem occurs when some power lines are too narrow, or an insufficient number of contacts or vias have been placed for the large current density carried.

Current density can be reduced by increasing the size of the metal lines or adding more contacts between metal lines. Adding more power lines on metal layers also reduces the current densities. In general, with more metal lines and vias used in the power distribution network, the electromigration failures are decreased while the IR drop is also reduced. In the early design planning stage, enough power lines should be provided in order to overcome the IR drop and electromigration problems.

The operating switching time (T_0) is defined as the minimum time between successive current switching operations, as shown in Figure 3-1. The current operating frequency is defined as $f_{sw} = 1/T_0$. The switching factor (s) is defined as a fraction of operating cycles over the life of the product during which a given circuit

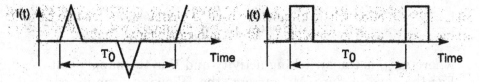

Figure 3-1. Switching time period.

switches. The average DC current (i_{dc}) is calculated based on the following equation:

$$i_{dc} = \frac{S}{T_0} \int_0^{T_0} i(t)dt \qquad (3\text{-}2)$$

In addition, two more current measurements are used for the EM analysis: RMS current and peak current. The RMS current is calculated as follows:

$$i_{rms} = \sqrt{\frac{S}{T_0} \int_0^{T_0} i^2(t)dt} \qquad (3\text{-}3)$$

where $i(t)$ is the current waveform, as shown in Figure 3-1.

The peak current (i_{peak}) is represented as follows:

$$i_{peak} = \max[\,|i(t)|\,] \qquad (3\text{-}4)$$

In the process design manual, the EM rules are specified to protect against two types of current-density-introduced failures: the standard EM and local heating EM. For the standard EM check, the rules define the maximum DC current I_{dc}, which is the function of the metal width, such that: $i_{dc} < I_{dc}$ for every metal line in the layout.

For the local-heating-enhanced EM, the rules define the maximum RMS current I_{rms}, such that $i_{rms} < I_{rms}$; and in addition, the maximum peak current I_{peak} is specified such that $i_{peak} < I_{peak}$. In the design, for any currents over the metal lines, the above EM conditions have to be satisfied: $i_{dc} < I_{dc}$, $i_{rms} < I_{rms}$, and $i_{peak} < I_{peak}$.

The DC average current limit I_{dc} can be translated into the maximum load capacitance allowed for the drivers in order to generate the current $i_{dc} < I_{dc}$. For the typical CMOS situation, where circuits are used to charge and discharge capacitances, the following formula may be used to translate I_{dc} limits into the capacitance limits [64]:

$$C_{\max} = \frac{I_{dc}}{s \cdot f_{sw} \cdot V_{dd}} \qquad (3\text{-}5)$$

where sxf_{sw} is the switching activity and V is the supply voltage. For the case of pure AC current, the following formula can be used to translate I_{rms} limits into the capacitance limits [64]:

$$C_{\max} = \frac{I_{rms}}{f_{sw} \cdot V_{dd}} \, \theta^{-1} \qquad (3\text{-}6)$$

where θ is defined differently for square, triangular, and sinusoidal waveforms based on the switching activity and clock cycle time.

Table 3-1 shows the EM limits (the maximum allowable current rules) for an eight-metal-layers process, where W represents the drawn metal width of the metal line, and 0.04 is the process shift for the metal width correction after manufacturing; that means that $W - 0.04$ is the actual or effective width of the metal line after manufacturing [64].

In the case of narrow strips where a single via or contact is permitted along the width, the general rules can be applied by using two or more contacts or vias along the line length. The general rules can be applied for the cases of wide lines, provided the maximum number of contacts or vias allowed along the width are used.

For a wide line crossing a wide line, the general rule can be applied by using the maximum number of contacts or vias to create an L-shaped array, as shown in Figure 3-2 [64]. Use of redundant vias is recommended where possible.

Table 3-1. EM current limits ($T = 105°C$) [64].

Metal Level	I_{dc} (mA)	I_{rms} (mA)	I_{peak} (mA)
M1	$4.05 \cdot (W - 0.04)$	$\sqrt{[235.8 \cdot (W - 0.04)] \cdot [(W - 0.04) + 0.704]}$	$100 \cdot I_{dc}$
M2	$3.30 \cdot (W - 0.04)$	$\sqrt{[96.1 \cdot (W - 0.04)] \cdot [(W - 0.04) + 1.408]}$	$100 \cdot I_{dc}$
M3	$4.80 \cdot (W - 0.04)$	$\sqrt{[95.1 \cdot (W - 0.04)] \cdot [(W - 0.04) + 2.068]}$	$100 \cdot I_{dc}$
M4	$4.80 \cdot (W - 0.04)$	$\sqrt{[69.9 \cdot (W - 0.04)] \cdot [(W - 0.04) + 2.816]}$	$100 \cdot I_{dc}$
M5	$7.05 \cdot (W - 0.04)$	$\sqrt{[81.1 \cdot (W - 0.04)] \cdot [(W - 0.04) + 3.564]}$	$100 \cdot I_{dc}$
M6	$7.05 \cdot (W - 0.04)$	$\sqrt{[63.2 \cdot (W - 0.04)] \cdot [(W - 0.04) + 4.576]}$	$100 \cdot I_{dc}$
M7	$7.05 \cdot (W - 0.04)$	$\sqrt{[51.7 \cdot (W - 0.04)] \cdot [(W - 0.04) + 5.588]}$	$100 \cdot I_{dc}$
M8	$7.05 \cdot (W - 0.04)$	$\sqrt{[43.8 \cdot (W - 0.04)] \cdot [(W - 0.04) + 6.600]}$	$100 \cdot I_{dc}$

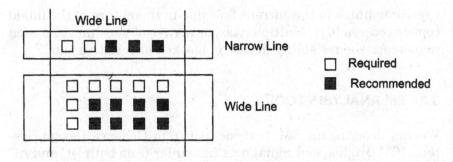

Figure 3-2. Via array for wide metal lines [64].

The maximum current allowed through all contact and via interfaces is described as follows. The number of contacts and vias placed across a line, perpendicular to the direction of the current flow, must be maximized or increased as soon as the line width permits, per layout rule restrictions, as shown in Figure 3-3.

If multiple vias are used, the allowable current value equals the allowable current per via times the number of vias. In all cases, the total current must not exceed the interconnecting metal line current limit, as shown in Table 3-1.

Multiple vias, or maximum coverage arrays of vias, added down the metal strip in the direction of the current flow do not increase the maximum current flow. Only the first via, or row of the via ar-

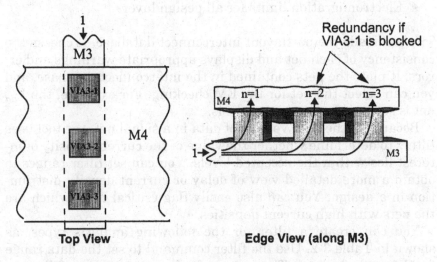

Figure 3-3. Reliability enhancement for placing multiple vias [64].

ray, contributes to the current flow due to the nature of the inlaid copper process [64]. Multiple vias, or arrays of vias, must be used to increase the reliability in case of blocked or resistive vias.

3.2   EM ANALYSIS TOOL

We will describe an EM analysis tool from Cadence Design Systems [65]. High-speed signal nets can suffer from both DC and AC electromigration problems. The tool uses two separate algorithms to provide comprehensive electromigration verification for any signal net. The tool can check nets in large designs without reducing the data, so it produces accurate results. It is typically used in high-speed clocks and data nets. It can highlight the areas of concern by using detailed simulation specifically designed to locate the electromigration issues.

The tool can produce the graphical output that clearly identifies the interconnect metals and vias of concern that violate EM rules. The tool uses two programs: one program accesses the design device information, and another program checks for signal electromigration. The tool requires the following inputs:

- An interconnect database
- Device capacitance data
- Driver-strength database
- Electromigration limits for all design layers

The tool loads the postlayout interconnect database. It checks the consistency of each net and displays appropriate warnings and errors. It plots the nets contained in the interconnect database, and you can select the net for the EM checking. For example, the V_{dd} net is selected for further analysis.

Because of the high volume of data in a signal net, the tool uses filters to determine whether the value of the current density of interest lies within the accepted levels. You can set filter ranges to obtain a more detailed view of delay or current density distribution in a design. You can also easily flag critical nets, which are the nets with high current densities.

You can create a filter for the following analysis types, as shown in Table 3-2. Use the filter command to set the data range for the analysis types. The syntax of this command is as follows:

Table 3-2. Current density analysis types [65]

Analysis Type	Symbol
RMS current density	Jrms
Average current density	Javg
Peak current density	Jpeak
Electromigration risk in each signal resistor for the signal net	Emrisk
Electromigration risk in a net	Emrisknet

Filter   [*jrms* | *javg* | *jpeak* | *jrec* | *emrisk* | *emrisknet*]   [*auto* | *range*]
[*on* | *off*] | *range min_value max_value*]

To set filter 4 for *emrisk* to be in the range of 10 to 50, enter the
following command:

>> *filter emrisk 4 10 50*

There are several methods used in the EM analysis by this tool
[65]:

- Method 1: calculating worst-case values without driver infor-
  mation.
- Method 2: calculating worst-case values with driver informa-
  tion.
- Method 3: calculating realistic values for Javg, Jpeak and
  Jrms.
- Method 4: calculating Javg, Jrms and Jpeak by using user-
  provided device current data.

Method 1 is the fastest but least accurate method. Using this
method causes the tool to overestimate the current in the net.
Start by calculating the worst-case values for peak, average, and
RMS current density, without the driver information. The tool
can apply the electromigration analysis to all signal nets in a
large design and filter out critical nets with potentially high current densities. This analysis can drastically reduce the number of
signal nets requiring further investigation.

The tool assumes that all inputs and bi-directional ports on a
net drive the net in parallel. For each driver, the tool assumes the
maximum driving strength defined by the default driver strength
and default port driver strength environment variables, as well as

a step voltage function at the driver inputs. For example, the default value for both variables is 10 Ω.

You must set the voltage range and cycle time by using the activity command. You can improve the quality of the estimate by adjusting the activity ratio on a per-net basis over consecutive analyses. The tool will report the nets that cannot be passed in the current density check.

Method 2 produces more accurate currents in the net than Method 1 and is almost as fast as Method 1. It requires the driver information—the direction and strength of the ports driving a net—to make more realistic current estimates. The tool will calculate the driver data and place it into a file. This type of analysis uses the same algorithm as Method 1, which enables you to repeat the electromigration checks for nets that failed in Method 1, calculating worst-case values without driver information. When you specify Method 2, you must use the Load Driver command to load driver strength information.

Method 3 is the slowest but most accurate method. Consider using this method only for critical nets, that is, nets that allow failures during the electromigration analysis with Methods 1 or 2.

Using the driver information, the tool uses a simulation method to determine J_{avg}, J_{peak}, and J_{rms} in every resistor. This analysis gives the most accurate results for each resistor in the net but requires a longer run time compared to Methods 1 and 2. Method 3 will require the driver information—the direction and strength of the ports driving a net—to make a more realistic current estimation.

When you specify Method 3, you must use the Load Driver command to load driver strength information. But more accurate analysis using the detailed simulation in Method 3 will increase the run time. Method 3 only analyzes the nets that failed in Method 2.

Method 4 performs the electromigration analysis by using precalculated average, RMS, and peak device currents. It can also define groups of devices that either charge or discharge a net. This methodology assumes that truly parallel devices, which are transistors with the drain, gate, source, and bulk connected to the same node, act together as a unit. It derives a separate solution for each driver charging or discharging the net.

For devices with no current specified, it assumes a zero current and does not calculate a separate solution. If you do not specify a current for any of the devices connected to the net, the tool issues a warning and performs no analysis. For each transistor, you can specify two average values as follows:

1. I_{avg_ds}: the average current flowing from drain to source
2. I_{avg_sd}: the average current flowing from source to drain

For each charging current, which is provided by a single device or a group of parallel devices, the tool calculates the average current by using the charging current and the capacitance of the net.

Another commonly used EM analysis tool, called RailMill, from Synopsys, Inc., is described as follows [5]. It simulates the power network of the IC design for EM violations. It will display a color-coded picture of the circuit showing the current densities in various areas. Red color indicates that the current density or electro-migration limit has been violated. Brown and orange colors are used for areas in which the values are quite close to the limits. The yellow portion of the circuit is where the current density value is one-half of the limit. Finally, the blue, green, and grey colors correspond to the much lower current densities.

The analysis tool separates the power network from the transistors by extracting a model of that network from the design layout file [5]. It performs transistor-level simulation of the IC to determine the current in each part of the circuit at each instant. An input vector set that reflects the operational behavior of the chip is used for the transistor-level simulation, so the power network will be simulated using the realistic currents.

Once the transistor-level simulation is completed, the calculated transistor current and the power network model serve to determine where electromigration problems exist. A graphical environment is provided with which users may perform iterative what-if analysis [5]. The user may make tentative changes as annotations to the power network, simulate and analyze them, and then display problems; designers may change the width of specific wires, add more power pad connections, add power lines, and delete power lines. All the tentative changes will not make real changes to the layout.

3.3 FULL-CHIP EM METHODOLOGY

Full-chip reliability has become more critical because advances in technology are yielding narrower interconnect structures and high-frequency designs [4]. This combination increases the risk of electromigration and joule heating failures in designs.

Traditionally, designers are given simple layout design rules based on the wire current density limits to which they must adhere. These limits, set to provide reliability over a broad range of circuit configurations, can make high-speed designs excessively large or impossible to design. This indicates that a methodology for the reliability budgeting is needed to permit engineering trade-offs between performance, design size, and lifetime.

This methodology must analyze the circuit to obtain realistic estimates of actual currents flowing in the circuit; apply advanced electromigration models to wire segments, usually based on Black's equation; and perform statistical analysis over the wires in the design to estimate the probability of the chip operating properly over its lifetime.

Due to the complex power grid and distributed blocks in a design, current flow from a chip pin to the gates cannot be determined without full-chip analysis. This is one of the reasons why full-chip electromigration analysis finds design problems. The current flowing in the chip may be taking a completely unexpected route through failure-prone portions of the power grid.

A design methodology includes extraction of chip interconnect data. It uses a static or dynamic full-chip analysis to determine current loading characteristics at the various device contacts to the power grid, and modeling mechanisms to report either wire segments likely to fail or overall chip lifetime characteristics.

Full-chip reliability analysis is part of the power distribution verification process and can be carried out in parallel with the *IR* drop analysis. The ability to apply reliability analysis at the full-chip level makes it possible to bring product reliability and reliability budgeting into the design cycles. The power grid electromigration analyses require the creation of models for a chip.

Model data is provided for each metal and via layer in the chip. Each metal-layer model provides the layer thickness and current density limits for peak, average, and RMS currents through wire segments. Different foundries provide different rules for threshold checks. Different model parameters may be applied for narrow wires and wide wires; an additional model parameter defines the boundary between narrow and wide wires.

Each via and contact model provides the current limits for peak, average, and RMS currents through each via/contact for threshold checks. A more detailed analysis of the reliability is made by calculating the theoretical time to the failure and EM-risk value for each segment, and using the proper failure statistics to obtain a failure probability as a function of time for the entire chip. The results are highly dependent on the choice of the statistical model used.

When the wire segments with the highest EM risk are identified, these can be provided to the designer for an engineering change order (ECO) if the overall chip probability is below specification. Improving the reliability of the latest reliable elements in the design will drastically increase the overall MTF of the design.

To fix the electromigration problems, the metal lines are widened while observing the possible warnings of electromigration failures. In addition, more vias and contacts are needed between these wider metals lines between different layers. Figure 3-2 illustrates this design concept.

3.4 SUMMARY

The power grid of a chip is operated primarily in a pulsed DC sense with respect to the electromigration analysis. Therefore, the average current data through the circuit is used to perform electromigration analysis on the grid. The full-chip transistor analysis tool will provide the average current drawn by each transistor connected to the power grid.

Each power grid is modeled with the voltage sources at the V_{dd} and V_{ss} pins, and the transistor tap currents at the device connection points. The large linear system is then solved to determine the precise current flowing through every wire segment and via in the chip. Once each wire segment current density has been determined, simple checks are applied to identify those wires in the design that exceed the thresholds.

4

IR VOLTAGE DROP

A combination of factors cause increases in *IR* drop failure. In the past, designers of low-frequency circuits implementing 0.35 μm three-layer metal processes rarely encountered *IR* drop or electromigration issues. However, designs with frequencies above 100 MHz, 0.25 μm processes, or four or more layers of metal increase the risk of problems. The *IR* drop problem is the voltage drop across the power grid due to the currents flowing through the power metal lines or metal resistances.

Lower metal resistance or smaller current definitely help solve the *IR* drop problem, but this may not be the case, due to the scaled-down metal pitch and increased power consumption. In addition, the tolerance of the *IR* drop decreases due to the lower supply voltage. Therefore, we need to address the *IR* drop problem in the power grid design.

This chapter is partitioned into six sections. Section 4-1 describes the causes of the *IR* drop in the deep-submicron chip. Section 4-2 gives an overview of the *IR* drop analysis. Section 4-3 describes a static *IR* drop analysis method [51]. Section 4-4 describes a dynamic *IR* drop analysis method [51]. Section 4-5 discusses circuit analysis with the *IR* drop impacts to improve the accuracy. Section 4-6 summarizes this chapter.

4.1 CAUSES OF *IR* DROP

The first set of causes is related to the advances in process technology. Chip feature sizes are decreasing in accordance with

Moore's famous law. Transistor sizes are decreasing to permit high-density designs. Transistors require a lower power supply voltage to avoid device failures. A lower supply voltage means that lower noise margins or smaller *IR* drops are permitted on the power grid.

On the other hand, the ability to design increasingly complex chips leads to increases in overall size and power dissipation. To design larger chips, more metal layers are being used to implement longer signal and power routing. Narrower wires have higher resistance than those used in previous technologies.

These higher-resistance wires and higher overall power currents lead to increases in *IR* drop or power switching noise. The conflicting design trend toward lower noise margins means that you must achieve a balance between the inherent power grid noise and the power supply noise margin to achieve a successful design.

The natural response to balancing the technology trends is to be more conservative in power grid design by adding more power lines on the chip layout to reduce the *IR* drop. But a more conservative power grid design means sacrificing the chip area, potentially a high cost. Other trends in processing technology present additional problems related to the *IR* drop. Via and contact resistances are not scaling in accordance with the transistor scaling. The trend is for them to remain the same or increase in metal resistance.

The parallel nature of data, such as that in a 64-bit wide bus, usually will place the drivers of each bit of the bus together or near each other. Large drivers in a local area are a common cause of the local *IR* drop problem. When the drivers of all those bus bits are switched in the same time window, a local *IR* drop will cause logic errors in the circuit.

The clock net in the chip must operate synchronously. Simultaneous clock switching introduces a large, instantaneous *IR* drop on the power grid. The clocks on some microprocessor chips consume up to 40% of the chip's total power.

In addition to the clocks, most circuit activity in a design occurs just beyond the edge of the clock, due to the higher frequency, creating a high instantaneous power demand after the clock edge. The overconservative design for timing will also cause IR drop problems. For example, oversized buffers are usually used in the critical speed paths, increasing power consumption. Conservative design for the timing must be balanced with the power grid optimization.

The location and design of I/O pads are a further source of the *IR* drop. Simultaneously switching output pads, which always have a large load, creates a strong demand for the power current and causes *IR* drop. The placement of I/O pads and power pins is a difficult design challenge. I/O rings normally have independent power rings and pads to prevent I/O ring *IR* drops from affecting the internal chip power.

Another common source of IR drop problems is the isolation of block power grids. It is common to isolate the power grids for sensitive blocks in the design, such as phase-locked loops and memories. However, power grid problems can result from excessive isolation or insufficient isolation.

Excessive isolation occurs when the block's power grid is so well isolated that the resistance from the power pad to the block is excessive, causing the *IR* drop. Insufficient isolation occurs when neighboring blocks create an *IR* drop that will impact the sensitive block. The *IR* drop in the sense amplifier is of particular concern for the memory design.

Many low-power design methodologies apply techniques to reduce the average power dissipation of a block. Techniques such as gated clocking isolate the power demands to the times of the block activity. Low power consumption does not necessarily mean low *IR* drop. If we design the block power grid on the basis of average power consumption, undersized power buses will create *IR* drop problems.

The last source of *IR* drop problems is errors in connecting global power grids to block power grids. It is common to design the global and local power grids separately. The power grid is designed to attach the block power grid at a large number of points after the block is finally placed.

Either manual or automatic techniques are used to insert the vias in the design where the grids are to be connected. This process may cause the attached points to be missed, resulting in a large *IR* drop to a portion of this chip.

4.2 OVERVIEW OF *IR* ANALYSIS

Power grid analysis helps to identify weak spots in the power network. Weak spots are the lower supply voltages that result in excessive *IR* drop or ground bounce. A good power grid analysis tool

not only helps you find such weak spots, but also helps you understand what you must change to improve the weak spots. The *IR*
drop analysis tool VoltageStorm™ Transistor-Level PGS from Cadence Design Systems will perform this task [51].

It includes static, activity-based, and dynamic analyses. Power
grid analysis involves the extraction of power grid and netlist
data from your chip layout, followed by the analysis of the power
grid and netlist. The interface between circuit netlist analysis and
power grid analysis is implemented by using the tap currents.

In most cases, each tap current is a transistor current, but it
could emanate from a variety of elements. Tap currents are currents arising from the connection of transistors to the power grid.
Figure 4-1 shows a typical netlist analysis view of transistors connected to a power grid. Each transistor is modeled with a tap current, as shown in Figure 4-1(b).

If the netlist has a million transistors connected to the V_{dd}
wire, data for a million transistors is passed to the power grid

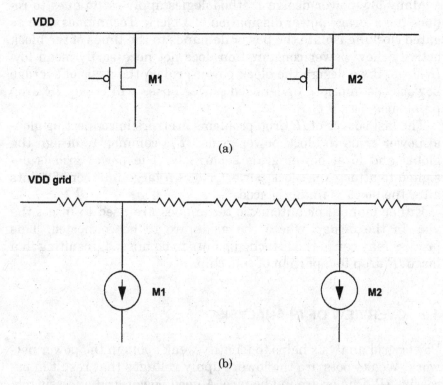

Figure 4-1. Tap current model of each transistor tied to V_{dd} [51].

analysis. The power grid analysis includes no information about any transistors other than those connected to the specific power grid being analyzed.

In a typical digital circuit design, one-third of the total number of transistors is connected to V_{dd}, one-third is connected to V_{ss}, and the rest are connected to internal nodes between logic gates. Because the primary elements in common between netlist analysis and power grid analysis are the transistors connected to the power grid, the power grid analysis models the transistor tap currents as the current sources attached to the power grid.

The tap current data file provides the details for each current source. Tap current files can be static—only a single current value is provided for each transistor—or dynamic—a sequence of data points is provided for each transistor. These currents are used to perform either a simple steady-state analysis or a dynamic analysis of the power grid.

Transistors have four terminals: drain (D), gate (G), source (S), and bulk (B). A typical *p*-type transistor representation is shown in Figure 4-2. The dominant current in a transistor is I_{DS}, the current flowing into the drain through the transistor and out the source. In a *p*-type transistor, this current is typically negative. In the power grid analysis, we are interested not only in I_{DS}, but also in the total currents flowing from and to the power grid: I_S and I_B.

The total power current is the sum of these currents over all transistors. The total current sink to the V_{dd} line, in Figure 4-2, is the sum of I_S and I_B. I_S is the sum of several currents in the transistor, as follows:

$$I_S = -I_{DS} + I_{CSG} + I_{SB} \qquad (4\text{-}1)$$

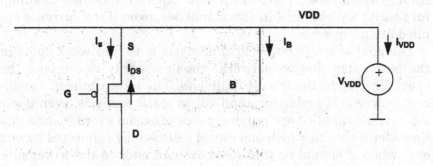

Figure 4-2. Tap current calculation [51].

where I_{CSG} is the current charging the transistor capacitance, C_{SG}; I_{SB} is the junction current including capacitive current between the source and the bulk; and I_B is the sum of several currents as follows:

$$I_B = -I_{SB} - I_{DB} + I_{CBG} \qquad (4\text{-}2)$$

where I_{CBG} is the current charging the transistor capacitance, C_{BG}; and I_{SB} and I_{DB} are the junction currents. I_D contributes to the total power dissipation for chips over a million transistors in size, but it is not a primary cause of *IR* drop.

In addition, the bulk current flows into either a well or the substrate of the chip and, therefore, usually introduces its load to the power grid at a location away from the transistor. For these reasons, we will consider only I_S in the power grid analysis, although the analysis also computes I_B during the circuit netlist simulation [51].

The following sections will compare static and dynamic analyses with the power grid analysis tool and show how static analysis can find problems in the power grid [51]. When it is used effectively and interpreted properly, static analysis with the tool can even find data-dependent power grid problems. We perform the static power analysis when the analysis of a power grid is based on the steady-state current modeling of the tap currents [51].

If we simulate the chip with thousands of test vectors and track the average current through each transistor connected to V_{dd}, we can obtain a long-term average behavior of the V_{dd} distribution network in the power grid analysis [51].

The challenge in static power grid analysis is obtaining sufficiently representative tap currents in a small computation time. An important lesson learned through experience is that meaningful results are obtained in static analysis, even if the currents applied are not precise.

The goal of static power grid analysis is to find weak spots in the power distribution network, not necessarily to compute the exact *IR* drop to the closest millivolts. The most common significant power grid problems stand out in static analysis, even if the tap currents applied are rough guesses of actual average currents. Consider a chip in which one row of cells is only connected on one end, when it should be connected on both ends to the power network. The result is that the *IR* drop at one end of the row is much larger than in all other rows in the chip. Even if the total power distribution of the chip is unknown, a specific row standing out above the others is a strong indication of a weak spot.

As another example, consider a set of drivers of a long bus, all powered from a specific location on the power grid. In this case, an *IR* drop failure may be data-dependent. However, in the static power grid analysis each driver is modeled by a larger current because of either the larger load on the driver or the larger transistors in the driver.

These larger currents in the static analysis highlight the weak spot without requiring you to simulate the specific vector to activate all drivers at once. You can still find problems without performing a significant amount of simulation.

If the currents are overestimated by the static analysis, which uses the worst-case switching activity, the method can provide current scaling information to the static analysis. For example, memory cells have substantially lower activity levels than the other circuitry, so dedicated current scaling factors are applied to these low-switching-activity regions.

The average currents assume equal amounts of rising and falling transitions on nets, so you can ignore currents due to the Miller capacitances in the transistors. The most significant advantage of static power grid analysis is that the requirements for extraction and netlist analyses are much lower, so we can rapidly perform the static power grid analysis.

Then we can apply a more extensive dynamic analysis while waiting for the chip to return, if the schedule does not permit it before the tapeout. It is recommended that one should always begin with static analysis before proceeding to dynamic analysis.

Dynamic power grid analysis uses simulation vectors to simulate the chip to obtain a finer solution of the chip's behavior. Although static analysis is quite effective in finding weak spots in the power grid, we may want to go to the next level of depth in analyzing the power grid. The dynamic analysis helps to identify false warnings caused by the temporal variation of currents.

Figures 4-3(a) and (b) show the current distribution based on the timing diagram. Obviously, the static analysis may treat the total current consumption the same for the current distributions shown in Figures 4-3(a) and (b), but the real *IR* drop will be much smaller if we can identify the current pulses in the timing diagram, based on the dynamic analysis for asynchronous transistor currents.

Each part in Figure 4-3 shows the current waveforms for transistors M1–M6 over a clock cycle. Each transistor has the same

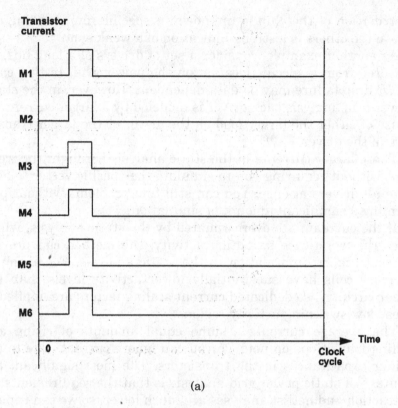

Figure 4-3. Current pulses in different timing patterns [51].

current pulse. The difference between Figures 4-3(a) and (b) is in the timing of the pulses. In Figure 4-3(a), all pulses occur at once, and in Figure 4-3(b) they are spread out over the clock cycle. Both sets of current waveforms yield the same average currents for all transistors. Depending on the specific characteristic of your chip design, the case shown in Figure 4-3(a) has a worse *IR* drop than the case shown in Figure 4-3(b).

You are likely to use dynamic analysis for one of the following four specific reasons [51]:

1. To simulate a specific test vector
2. To identify which specific test vectors activated an implementation weakness
3. To examine the time correlation of tap currents
4. To obtain a better estimate of the realistic magnitude of *IR* drop

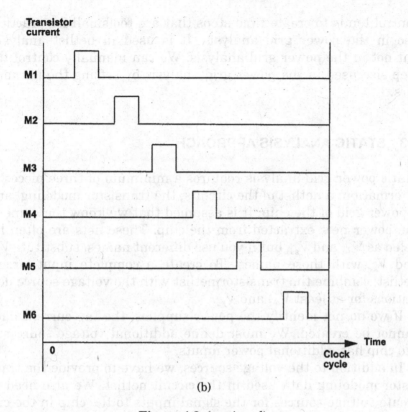

(b)

Figure 4-3 *(continued)*.

The simulation of a specific test vector is common in memory design to test power grid behavior under specific corner cases. It is also used when worst-case *IR* drop test vectors are known before analysis. Simulation to identify which specific test vector activates a weakness is useful when you cannot change your power grid, but you can change your power profile by changing the vectors using microcodes.

. Examining the time correlation of tap currents is a valuable check to avoid the static averaging current issue. A better estimate of the magnitude of the *IR* drop is used when the cost of fixing a weak spot is high and you want a more precise analysis before making the decision [51].

Dynamic power grid analysis is a type of transient analysis. Transient analysis assumes the application of automatic timestep control. However, performing full-chip netlist and power grid analyses require many resources. Automatic timestep control tends to create time steps that are too small for practical use in the power grid analysis. It is used in netlist analysis, but not in the power grid analysis. We can manually control the step size used in the power grid analysis by setting the parameters.

4.3   STATIC ANALYSIS APPROACH

Static power grid analysis requires a minimum of three pieces of information: a netlist of the circuits, the transistor modeling, and a power grid of the chip. It is assumed that we know the name of the power nets extracted from the chip. These nets are often labeled as V_{dd} and V_{ss}, but if you use different names, substitute V_{dd} and V_{ss} with those names. To create a complete input circuit netlist, combine the transistor netlist with the voltage source definitions for at least V_{dd} and V_{ss}.

If we do not identify the power supplies, the tap current data cannot be created. We must define additional voltage sources if the chip has additional power inputs.

In addition to the voltage sources, we have to provide the transistor modeling data used in the circuit netlist. We also need to create voltage sources for the signal inputs to the chip in the circuit netlist. The signal voltage sources are required to be piecewise linear input sources with a single initial voltage.

If we are going to apply vector-based simulation, the data in the vector file overrides the piecewise linear data. For static power grid analysis, the power grid extraction from the chip must only contain the resistances. In the steady-state analysis, inductances and capacitances are treated as shorts and opens, respectively, so the extraction time is reduced by considering only the metal resistances in the power grid.

If you intend to analyze both V_{dd} and V_{ss}, extract them individually and do not extract both into a single power grid database. We know where the power pins are located in the power grid. Voltage sources are defined in the power grid analysis for the power input pins at the locations.

A different voltage source is defined for each pad and can include a series of resistances and inductances. Different sources are used because each has different behaviors resulting from the characteristics of the power grid in operation.

The passing of tap currents from the circuit netlist analysis to the power grid analysis uses the transistor names for identification. It is therefore critical that the transistor names be consistent between the circuit netlist and the power grid. The power network and the circuit netlist use the same extractor from the layout. For the static power grid analysis, neither net names nor transistor names need to match any schematic that you may have.

Schematic net names are only required if you supply activity data. One of challenges in the static power grid analysis is to obtain an accurate estimate of the distribution of the power consumption in the chip. A rough power consumption estimate method has been developed [51]. It uses various forms of data to derive the distribution of the power consumption in the chip. It requires a default chip frequency and the following optional information. The specific clock inputs and the clock frequencies are used to trace the clock domains in the design.

Any portion of the design not assigned with a specific domain is assigned the default chip frequency. The gates on the clock distribution network are modeled as operating at the specific clock frequency. We can derive an activity rate for the logic circuit that is not on the clock tree, based on the clocked domains [51].

We can also specify the activity rates or frequencies of the specific nets in the design. This information is used to set the known activity rates of specific nets in the design. It is propagated forward and backward, considering the logic functionality to improve the estimation of the activity rates of nearby logic circuits.

We can further specify the power consumption of specific blocks in the chip. When the actual power consumption of a specific block is known, this number is used to automatically scale the distribution of currents in the block, so that the total estimated power consumption of the block will match the specified one.

We could specify the power consumption of the entire design. Once we estimate the various portions of the chip to determine their power distribution, the specified total power consumption of the chip is used to scale the estimated currents in the design to match the specified one.

The following sections explain the power estimation based on the maximum saturation currents. No capacitance or vectors are required, so the turnaround time is that of the connectivity and resistance extraction.

To use this method, we should have an estimate of the total power dissipation of the design in the form of average current. Additional information of the block power consumption also improves the quality of the analysis. The peak saturation current, I_{DS}, for each transistor connected to the power grid, is calculated based on the device's IV curve and the transistor sizes. This peak saturation current is used as the tap current.

Parameters to scale the V_{GS} and V_{DS} voltages are applied to compute the saturation current. In addition, certain transistor configurations result in no I_{DS} current, because transistors with shorted source and drain or gates have been turned off.

Although saturation currents may seem to be an inaccurate method for deriving average currents, they have been quite successful in finding weaknesses in power grids. We can scale the saturation currents by the scaling factors, based on the specified power consumption of blocks. If we know the specific power dissipation of the blocks or chips, we can scale the currents accordingly.

We can also use net activity data to estimate the power for the power grid analysis. The clock is defined as having an activity ratio of 1.0. The net activity is used in conjunction with the net capacitance, V_{dd} voltage, and chip frequency to derive the average current of the transistors connected to the power grid.

Given these parameters, the average current consumed by a gate is derived from the following equation:

$$I_{AVG} = A \cdot C_{GATE} \cdot V_{dd} \cdot F \qquad (4\text{-}3)$$

where A is the activity ratio of the gate, C_{GATE} is the total capacitance of the nets in the gate including the load capacitance, V_{dd} is the supply voltage, and F is the chip frequency.

Computing tap current on the basis of net activity introduces two additional requirements for the layout extraction: (1) parasitic capacitances must now be computed for signal nets, and (2) back-annotation of the net names from the schematic.

We can also derive the average transistor currents by performing vector-based simulation in the netlist analysis. This can achieve more accurate average power grid currents by using the transistor-level simulation of several vectors. This approach is most commonly used at the block level for electromigration analysis.

The tool uses one test vector input file, performs the simulation over the vectors provided, and tracks the tap currents [51].

It tracks the average, peak, and RMS currents at once and reports them in three separate tap current files. Each tape of tap current data provides a different perspective of simulation behaviour, allowing us to select which is the best suited to the need.

Computing the average tap current on the basis of the vector simulation requires one more additional condition for the layout *RC* extraction: the parasitic capacitances should now be computed for signal nets. If the vector input signals are not labeled in the GDSII input, you must back-annotate the signal names in the schematic to the extracted netlist from the layout.

4.4   DYNAMIC ANALYSIS APPROACH

Dynamic power grid analysis is the next step to improve the tap current estimation accuracy, based on the input vectors at I/O pins. It also includes the time variation of the currents in the analysis. Rather than averaging the currents as in the static power analysis, this dynamic power analysis enables us to see the fine time variation of currents over a clock cycle.

The challenge in the dynamic power analysis is to find the weakness in the power grid by using the minimal amount of computational time. A technique in the dynamic analysis includes the capability for a form of vector compression in the creation of the dynamic tap current data [51].

The vector compression is intended to create an effective worst-case *IR* drop test vector by merging the behavior of many vectors into a single equivalent vector set. The dynamic power grid analysis introduces two additional requirements for the extraction beyond those of static analysis as follows:

1. The parasitic capacitances must be computed for both signal nets and power nets, which are merged into the power grid for analysis.
2. In addition, if the vector input signals are not labeled in the GDSII input, we need to back-annotate signal names to the layout netlist extracted.

The capacitance on the power grid is due to two sources: parasitic capacitances and transistor capacitances. Parasitic capacitances are generated from the RC extraction from the layout; and the transistor capacitances are embedded in the dynamic tap currents extraction. The decoupling capacitances are also included in the transistor capacitances.

The dynamic power analysis processes the dynamic current data as piecewise constant current sources. The recommended step size is about a single gate delay. Another criterion is to use one-tenth of the clock cycle as the step size, so if our clock cycle is 10 ns, it will use 1 ns as the step size in the dynamic simulation.

If we want to include the pin inductance, a smaller step size is required, such as one-twentieth of the clock cycle. The power grid solution is performed by constructing and solving the massive matrix problem. The size of the matrix describing the resistive connectivity of a full-chip V_{dd} network can be very large.

The number of resistors in the V_{dd} network can be the number of metal layers times the number of transistors in the circuit. In the five-to-six metal layers process, the ratio will be five to six times; and 10 million transistors will have 50 million resistors in the network. The matrix to solve the power grid analysis is huge.

VoltageStorm™ from Cadence Design Systems uses vector compression to reduce the overall computational time, because the time to solve a large matrix for each of the large number of time points can be very large [51]. If we simulate the chip for 100 vectors, and select 10 steps per clock cycle in the dynamic analysis, we may perform 1000 solutions of the power grid. This may not be practical with existing computational resources. The vector compression reduces the number of solutions to 10. It is useful when our objective is to resolve the temporal issues of the static analysis or to estimate the magnitude of worst-case the *IR* drop more precisely.

The dynamic analysis will introduce the time correlation to the analysis data. The chips are synchronous in their behavior, with the clock being the synchronous signal. Introducing the temporal correlation in the dynamic analysis splits the activity occurring at different portions in the clock cycle, rather than modeling the clock cycles as a single time-averaged value. The key is to improve the resolution in a clock cycle, not across many clock cycles or vectors [51].

For example, assume that we split a 10 ns clock cycle into 10 buckets {B1 – B10} of 1 ns each, B1–B10. B1 corresponds to the interval 0.0–1.0 ns into the clock cycle, B2 to the interval 1.0–2.0

ns, and so on. Figure 4-4 illustrates the current for gate G1 over a clock cycle [51].

If gate G1 can only switch in the time interval corresponding to bucket B2 in the dynamic analysis, the current value for gate G1 in buckets B1 and B3–B10 should be 0.0 A in all clock cycles. The current value in bucket B2 may be 0.0 A in some clock cycles and nonzero in others. Over the 100 vectors, 1000 total buckets correspond to gate G1. The 1000 buckets correspond to 100 vectors and B1–B10 offsets into each vector.

The second concept in vector compression is that of peak analysis, with a goal of finding the worst-case current. When simulating to determine the peak current of a transistor, we take the maximum value found for the transistor currents at each time point.

We would like to find the worst-case set of current buckets for gate G1 to create a current waveform for a single worst-case clock cycle. We want to find the peak over many vectors, but meanwhile want to preserve the time offsets or buckets in clock cycles.

In summary, the vector compression technique will assign the worst-case bucket with the largest current to the specific gate (e.g., G1) for many vectors and so on for all gates. For example, Table 4-1 shows the peak currents in different input vectors at bucket 2 for Gate 1, so 2.1 mA is used for the largest current for Gate 1.

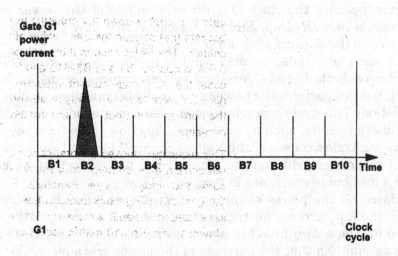

Figure 4-4. Current distribution to gate G1 [51].

Table 4-1. Current values in buckets for G1 [51]

Test Vector	B1	B2	B3
1	0	1.0 mA	0
2	0	2.1 mA	0
3	0	1.4 mA	0
4	0	0.2 mA	0
5	0	0.0 mA	0

After the assignment is done, we can assign the worst-case currents to tap currents. So the dynamic simulation is only done for the number of timing buckets in one clock cycle. Notice that the tap current assignment by many test vectors may be overestimated in this technique [51].

The computational time for the power grid analysis is only proportional to the number of buckets instead of the number of buckets multiplied by the number of input vectors. This processing independently takes place for each tap on the power grid.

If we have a V_{dd} power grid with 1 million transistors, we will have 1 million sets of buckets. Each bucket set is the compression over all the test vectors. Vector compression tries to synthesize a worst-case *IR* drop test vector. How many vectors are required to obtain a sufficient amount of data? The answer is a function of the chip and the vectors that we apply.

We probably can obtain high-quality results from as few as one vector, because the clock is a primary source of the power consumption and *IR* drop. Simulating one vector may give you insight into the performance of the power grid.

In some situations, we may want to perform power grid analysis for each clock cycle, avoiding vector compression technology [51]. We can perform the power grid analysis on a single vector in isolation. This cycle-by-cycle flow is sometimes used in mutually exclusive circuits, such as memories. This flow is useful when we have multiple licenses of the VoltageStorm™ tool and we want to split the power grid analysis into several pieces, so that we can use a number of machines in parallel [51].

Based on the pieces of the power grid, each piece is analyzed with the tap currents for transistors, and then we analyze the *IR* drop for the entire full-chip power grid. This method is based on the assumption that the currents of the power grid's pieces do not interact for different vectors [51].

4.5 CIRCUIT ANALYSIS WITH *IR* DROP IMPACTS

Theoretically, the tap currents of the circuit rely on the supply voltage of the power distribution network. Hence, the power grids' *IR* drop analysis and tap current analysis of the circuits interact with each other. But if we simulate them one by one at each time point, it may take a long computational time to improve the *IR* drop analysis accuracy. The power grid analysis creates an *IR* drop report that contains the voltages computed after the static power grid analysis.

With the tap current file, the identification device used to pass data back is the name of the transistor tap. We can repeat the netlist analysis by using the unique voltages for each transistor connected to the power grid.

The impact of the feedback is that the tap currents computed again will have a smaller magnitude than in the first pass, because the lower-power grid voltages reduce the voltage swing of the gates. The gate delay will also be larger.

But the gate speed should not have much impact on average currents unless the functionality is altered. In examining the results of two passes through the loop, we obtain both the worst-case and best-case *IR* drop results.

In the first pass, we can observe the worst-case *IR* drop, because the voltage on the power grid is highest using the ideal V_{dd} value for all nodes in the power grid. In the second pass, we could observe the optimistic *IR* drop, because the *IR* drop values fed back to the circuit netlist analysis will reduce all the tap currents.

The next step to improve in accuracy is to feed the dynamic analysis results into the circuit analysis. Time-varying power grid voltages alter the speed of the transistors to obtain the most accurate performance estimate of the design. In this methodology, dynamic *IR* drop data is fed back to the netlist analysis. The waveforms applied to each tap current of the transistor are now dynamic rather than static.

4.6 SUMMARY

The analysis of the power grid can be done either by static or dynamic methods. The static method uses the average current and current scaling factor to estimate the static number for the *IR* drop. It is faster and easier to identify the weak spots in the power distribution grid by using this method.

The dynamic method improves the accuracy by simulating the power grid and tap currents in multiple time points of the clock cycle, similar to the transient analysis of circuit simulation for both the circuit netlist and the power grid resistance network. The dynamic method is not usually practical for a full-chip scale due to the long simulation time required in multiple input test vectors, but it is worthwhile to try it out using one or two test vectors or the vector compression technique [51].

The best solution to a given *IR* drop problem depends on the type of the *IR* drop, the chip architecture, the chip layout, and the functionality. Several approaches can be used in the circuit and layout design to fix power drop problems as follows [51]:

- Widen metal lines.
- Add or remove straps to redirect the currents.
- Reduce the circuit sizes while meeting the performance targets.
- Add decoupling capacitors to the design.
- Use C4 or flip-chip technology.
- Add more V_{dd} pads to the design.
- Connect buffers to different power buses.

Using the power grid analysis tool VoltageStorm™ from Cadence Design Systems, we can make the ECO (engineering change order) for the power network design [51]. In addition to the analysis capability, the tool adds layout exploration capability, which enables the designer to perform power grid ECOs within the tool.

The designer can remove all the power grid problems from a design in a single ECO pass. Once the power grid is clean, we could create a single ECO list, called a change report, which guides the implementation of the layout modifications necessary to create the clean power grid design.

The exploration capability in the VoltageStorm™ tool enables us to quickly experiment with the power grid change, and then use the static power grid analysis to show the effects of these modifications on the impact of the power grid performance [51].

Because all the ECO changes are implemented within the framework of the tool, we do not need to reextract and reload the power grid network with each ECO, thus saving turnaround time.

5

POWER GRID ANALYSIS

This chapter will explain how to use CAD tools to help you find the weak spots in the power grid. We chose to use the VoltageStorm™ tool from Cadence Design Systems, Inc. although several other CAD tools perform similar tasks [51]. Weak spots are implementation characteristics that result in excessive IR drop, electromigration stress, or pin currents during the operation of the chip.

There are three approaches to finding weak spots. The first is finding the weaknesses in the power grid that are likely to impact the proper functioning of the chip, regardless of the magnitude of the impact. This approach is quite common and best addressed by using static analysis. It is strongly recommended to use static analysis before dynamic analysis, because static analysis can find the problems quickly.

The second approach to finding weak spots is to predict a worst-case IR drop vector on the basis of the limited coverage of the vectors for analysis.

The third approach to finding weak spots is to address the precise voltage drop on the grid for a specific test vector. This approach is common in memory design or when the cost of changing a design is high and we want to determine the exact magnitude of the IR drop.

This chapter is organized in six sections. Section 5-1 describes the data preparation and provides an overall introduction to a CAD tool used for the IR drop analysis. Section 5-2 explains the steps needed to execute the CAD tool. Section 5-3 discusses advanced static analysis, such as the activity-based analysis method. Section 5-4 discusses a dynamic analysis method that is similar to the transient analysis of the power network. Section 5-5 discusses layout exploration—changing the layout and then resubmitting the power grid analysis within a CAD framework. Section 5-6 summarizes this chapter.

5.1   INTRODUCTION

VoltageStorm™ uses several tools to perform power grid simulation for weak spot identification [51]. VoltageStorm™ uses Thunder, which is a netlist analysis tool, and Lightning, which is a power grid analysis tool [51]. Thunder performs a transistor-level analysis of the chip. It analyses the entire transistor netlist by using the voltage sources, transistor model data, and vectors that we provide. Notice that the power grid in Thunder is modeled as a single node.

Lightning performs a detailed analysis of the power grid node in which the node is represented by its resistor, inductor, and capacitor components. It only processes the devices connected directly to the power grid. The power grid is modeled as a linear circuit with voltage sources representing the power pins and current sources that represent the transistor taps connected to the power grid [51].

The power currents flow from the voltage sources through the grid and out the current taps. Proper analysis requires all three components: the voltage sources, the transistor–inductor–capacitor grid, and the tap current sinks [51]. Thunder calculates the current information for each device connected to the power grid (V_{dd} or V_{ss}) and passes these currents, plus device capacitances for dynamic power grid analysis, to Lightning. The interface between the tools is based on the names of the devices connected to the power node.

Thunder passes the current and capacitance data to Lightning for each transistor. Lightning passes the IR drop data to Thunder for each transistor. We must prepare a circuit netlist file. The power sources of the chip are defined in the netlist, which are used by the tool to identify the gates and to perform the simulation.

The primary inputs of the chip are also defined in the netlist to which the input vector files can be applied. These inputs should not be defined as DC sources because Thunder treats DC voltage sources as power sources to which vectors cannot be applied.

If the input voltage is a constant value over a specific simulation time, define the source as a piecewise linear (PWL) waveform at a single voltage. The output pin load of the chip has to be defined also. It is important to model the loading on the chip outputs in both activity-based analysis and vector-based simulation. A common error in the analysis and simulation is to forget the output pin loading in the circuit file.

In addition, the bidirectional pin loadings of the chip have to be defined, because they act like outputs during some time intervals. In the circuit netlist, we need to specify the link path to the transistor modeling data, which tells the tool how to compute transistor currents and capacitances as a function of the voltage.

We can refer to multiple sets of modeling data for different devices or place data for multiple models into a single directory. We also need to specify the link path to the transistor netlist and coordinate file used by the power grid analysis tool [51].

The netlist is usually hierarchical, although the flat netlist will be accepted in the circuit simulation tool. The coordinate database provides the geometric data about the locations of the devices in the layout, which is also used in the graphical output.

Finally, the circuit netlist can contain the path to the parasitic capacitance database of signal nets used and can be back-annotated into the circuit schematic node names for complete simulation with parasitic RC effects.

The power grid database is also required by the analysis tool. The power grid database contains the resistors and capacitors of the power net in the chip. We also need the locations of the power supplies that we want to model. When analyzing a block of the design, we select a number of locations on the periphery of the block where power will be connected to the block.

To model the package characteristics, we can define a series resistance as well as the inductance for the power pins. The accurate modeling of inductance requires smaller time steps for dynamic analysis.

Another database is the estimation of the transistor peak saturation currents, which is also called Ipeak analysis [51]. The methodology for the current estimation used in the static analysis is to estimate the average currents throughout the chip by computing the peak saturation currents for all the transistors connected to the power grid, followed by some simple scaling of the currents.

The above method is very simple and assumes that the average current of a transistor is somehow related to its size. Although this assumption is not strictly true, the results for a large number of transistors connected to the power grid highlight the problem areas of the power grid, if not their exact voltages.

Effectively using the circuit design experience for this analysis, and filtering and displaying data, will find most problems in the power grid design. It is claimed that accurate dynamic analysis of the power grids will show similar symptoms of those problems in static analysis using the above methodology [51].

5.2 EXECUTING THE TOOL

The following sections show the steps used to load the input databases, do the power grid analysis, and show the IR drop analysis [51]. The next section will show a more specific design example for the application of this CAD flow.

1. Move to the working directory in the UNIX shell:
   Shell>> *cd $thunder_working_directory*
2. Start the Thunder tool to load the netlist and compute the peak saturation current:
   Shell>> *thunder*
   Thunder> *load design.ckt*
   The above step is to load the circuit netlist file.
   Thunder> *pwrnet ipeak VDD*

The above step will compute the peak saturation currents. The command creates an output file named VDD.ipeak, which contains the desired peak currents. The currents are computed for each transistor connected to the V_{dd} voltage source, assuming a V_{GS} voltage magnitude of V_{dd} and a V_{DS} voltage magnitude of V_{dd}.

The current estimation is based on the IV curve and transistor size from the specific SPICE simulation deck in a specific process technology. We can also scale the resulting current to match the realistic average current on one design example, as shown in the next section.

Notice that the transistors tied to DC voltage sources,

which will turn these transistors off, are assigned 0.0 A current. We can exit the Thunder window as follows:

Thunder > *quit*

An alternative way of using Thunder is to create a command file, for example: ipeak.cmd, which contains the three Thunder commands introduced. Then we can use the command line version of Thunder to perform the analysis by entering this command as follows:

Shell > *thunder.tty ipeak.cmd*

The above command creates the *VDD.ipeak* output file, the same as the *pwrnet* command's output. Next, we can run Lightning by using the following steps, which will load the power grid *RC* network modeling, specify the power source pin locations, load the Ipeak current data file (*VDD.ipeak*) generated in the above steps, and then solve the linear network of the power grid modeling with *RC* and tap currents, and show the lowest voltage across the full-chip power grid.

3. Move to the working directory containing the V_{dd} power grid database:

   Shell>> *cd $lightning_working_directory*

4. Run Lightning:

   Shell>> *lightning*

   Lightning > *load design_VDD.mhdr*

   The above step loads the binary power grid database for V_{dd} and displays the power grid in the plotter window. The metal layers are shown in different colors—such as M3 in purple, M2 in tan and M1 in blue—in the layout display:

   Lightning > *putvsrc M3 Vsrc1 3.3 24000 17000*

   Lightning > *putvsrc M3 Vsrc2 3.3 284000 17000*

   Lightning > *putvsrc M3 Vsrc3 3.3 24000 12000*

   Lightning > *putvsrc M3 Vsrc4 3.3 284000 12000*

   The above step is used to define where the power source pins are placed. Four V_{dd} pads are placed in the chip boundary. M3 is the power line, which is started from the V_{dd} input pin. {24000 17000}, etc. are the X–Y locations of the pads in the layout with the drawn dimensions.

   Be sure to name each source differently. After each command, a white dot corresponding to the placement of the voltage source in the layout plotter window can be seen. The voltage is actually placed at the power grid subnode on M3 near the specified location. Units are in μm in general for X and Y coordinates, which are in the drawn layout sizes.

We can also place the voltage sources by using either a command file or the graphical user interface. One command file, called *vsrc.cmd*, can be created to contain the four *putvsrc* commands in the above step.

5. We now have the voltage sources placed on the power grid. We need the tap current model, which was calculated earlier and is stored in the file named VDD.ipeak. We load the currents in the Thunder working directory into the Lightning tool as follows:

   Lightning > *iload $thunder_working_directory/*
   VDD. ipeak

   As a matter of practice, after loading the static current data, use the following command:

   Lightning > *scan tc*

   The above command reports the statistics about the tap currents loaded. For example, the following result will show that the current range is 0.0 A to 0.004 A with an average current of 0.001 A and a total current of 17.3 A.

6. We have not scaled the current yet, and the above statistics are the sum of all transistor saturation peak currents connected to V_{dd}. In reality, not all transistors will switch at the same time, and the worst-case total current consumed by the circuit or drawn from V_{dd} will be smaller than the summation of all these transistor currents, depending on how many switching transistors occur in the worst-case application.

   However, to identify the switching patterns of transistors in the circuit to V_{dd} will require a long computational time using the dynamic simulation of the circuits in multiple timing steps.

   The methodology used in this static analysis is to roughly estimate the total current either by measurement or current simulation tool using multiple vectors, and then apply one *currentscalefactor* to the estimated peak current sum in VDD.ipeak by using the following command:

   Lightning > *setenv currentscalefactor 0.01*

   The above scaling factor is decided by designers and applied to the tap current scaling for later power grid analysis. We can recheck the static current load using the scan command again as follows:

   Lightning > *scan tc*

The statistics show that the total current is now 0.173 A, scaled by 0.01 from original 17.3 A. This 0.173 A current will be used in the linear circuit analysis of the power grid modeling.

7. We can perform the power grid solving to calculate the voltages across the power grid based on the tap currents, voltage sources, and resistive model of the power grid, which have been loaded in the above steps:

> Lightning > *solve*
> Lightning > *scan ir*

The *solve* command prints a number of messages, ending with the memory utilization of the solve command. The *scan ir* command scans the node voltages in the power grid and reports their range. For example, the node voltage range is 3.13 V to 3.3 V. The worst-case *IR* drop is about 0.17 V. It indicates the minimum voltage in the power grid.

The high volume of data processed by VoltageStorm™ makes normal reporting of every item in the database excessive in size, as well as nearly impossible to sort. Therefore, VoltageStorm™ uses a concept called filtering. For example, when screening for excessive *IR* drop, we may be interested in seeing where the *IR* drop exceeds 10% of the V_{dd} voltage in the power grid. Effective use of filtering on the various analysis types in VoltageStorm™ gives significant insight into the behavior of the chip.

Table 5-1 shows the most common analysis types supported by VoltageStorm™ [51]. Each analysis type can have up to eight fil-

Table 5-1. Analysis types in the VoltageStorm™ tool [51]

Analysis Type	Option	Abbreviation
Tap current (current drawn by transistors connected to the power grid)	*Tap_Current*	Tc
IR drop (voltage on each node of the power grid)	*IR_drop*	Ir
Resistor current (current through resistors)	*Resistor_Current*	Rc
Current density (current through a metal divided by the metal area)	*Current_Density*	Rj
Electromigration risk (probable time until failure because of electromigration)	*EM_risk*	Er
Resistor voltage (voltage drop across a resistor)	*Resistor_Voltage*	Rv

ters, and each filter contains a range. We can establish a set of filter ranges by using the following methods:

(a) Selecting the auto-filter setting on the command line
(b) Entering the range on the command line
(c) Using a command line
(d) Interactively selecting filters from the dialog box

Figure 5-1 shows a graphical representation of the ranges of filters. For example, node voltage 3.21 V is located in Range 4. We can set the *IR* drop filters both automatically and manually by using the *filter* command [51].

For example, the filters are set to assign the plot colors in such a way that the red color is assigned to the node voltages in the lowest range from 0.00 V to 3.14 V, the orange color to the voltages in the medium range from 3.14 V to 3.16 V, and the green color to the voltages in the high range from 3.16 V to 3.3 V.

Then, the *scan* command prints the range of voltages from the *IR* drop analysis with the number for each filter. The *plot* command will create a color-coded plot similar to the thermal plot. It

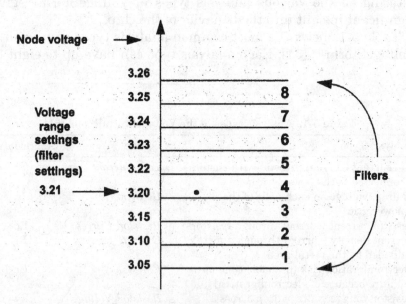

Figure 5-1. Filtering ranges of voltage data [51].

shows where the circuit has the largest *IR* drops, as well as the voltage trends from the power pins to each area of the chip.

A design example shows that the red colors in the plot, which has the largest *IR* drop, are located in the central control units on the left side of the chip [51]. The reason for this is that the power routing for much of the control circuitry is provided only from one side of the block, yielding high *IR* drops at the isolated end of the power bus, whereas the power is supplied to the top and bottom blocks from both sides of the block.

We can also use the VoltageStorm™ tool to view the geometric distribution of tap currents. We can use the *scan tc* command to view the total currents in the design. We can also create filters to plot the tap currents. Before doing this, we can set the analysis tape to *Tap_Current*, as shown in Table 5-1. One design example shows that the larger areas of currents are located in the data path units of the chip and smaller currents in the control units.

Resistor current value distribution has different characteristics than node voltages. The maximum current should be near to the power pins and the minimum near the transistors. After observing the *IR* drop in the plots, the next examination is of the current flows in the circuits to create the *IR* drop. The current flow trends may not be as you expected or currents from several power pins may merge in the middle of the chip to create a high current through the wires with high *IR* drops.

We can use *scan rc* command to get the data for the resistor current, and set the analysis type to *resistor_current* before the filtering and plotting. Finally, we can use the *plot rc* command to obtain colorful plots with the highest-current chip area in red, the medium resistor current area in orange, and the low-current area in green.

Once we understand how current flows through the chip, we examine the current densities of metal wires, which is a first-order indication of the electromigration failures within the chip. Use the *Current_Density* analysis type to examine them.

We can change the current density limits for metal layers. The current density reporting is not based on the actual value, but on the ratio of wire current density over the required limit. If the ratio is more than 1.0, there is a potential electromigration failure in that area.

Summarizing the modeling, analysis, and viewing results using the VoltageStorm™ (Thunder and Lightning) tool suite from Cadence, the recommended design flow is described as follows [51]:

1. Simulate the transistor current data of the chip using the Thunder tool.

2. Load the power grid into Lightning, and place the voltage source pins.

3. Use the *image* command to create a GIF image of the grid for reference.

4. Load the transistor current data from Thunder into Lightning.

5. For the blocks whose relative activity we know, use the *scalecurrents* command to scale the currents for the blocks. The scale factor for a memory is typically 1 divided by the number of words. If the design has blocks whose activities are exclusive, only some portions of the chip can be active at any one time. Scale the currents in those blocks accordingly.

   For example, if a block has four units, only one of which can be active at a time, scale the currents for that block by 0.25, or set the *scalecurrents* to 0.25 in this block.

6. Generate the plot of tap currents to visually inspect the areas of the scaled currents. If the scaled currents for some blocks were missed, go back to Step 5 to observe the reasonable tap currents for the power grid simulation. Once satisfied with the tap currents data, generate a GIF image of the tap currents plot for reference.

7. Given the total current reported by the tap currents scan and an estimate of the expected total power current, we can compute the appropriate value by setting the *CurrentScaleFactor* environment variable. It is important to use the *CurrentScaleFactor* environment variable to scale the total current consumption estimated by the tool to be matched with the one from real estimation or measurement.

   Notice that the real estimation or measurement of the power consumption for the chip is derived from the total power consumption during the active logic switching. We may not want to average in the inactive time of the design, because it may produce a substantially lower value than the average power consumption.

8. Use the *solve* command to solve the power grid I-V equations.

9. Use the *savestate* command to save the solved results. Saving the result will enable reuse later without having to

solve the power grid again, which will save the computational time for a large-size chip.

10. Iterate between using the *scan ir* command and setting the filter to derive a good set of filters to observe the *IR* drop. These filters are generally equal-sized steps. We can generate the plot of the *IR* drop by using the *plot ir* command. Use the *image command* to create a GIF image of the *IR* drop plot.

11. Iterate between using the *scan rc* command and setting the filter to derive a good set of filters to observe resistor current flow in the chip. These filters are generally decreased in magnitude logarithmically. We can generate the plot of the resistor current by using the *plot rc* command. Use the *image command* to create a GIF image of the resistor currents plot.

12. Create a plot for each layer with only the metal layer and its error layer turned on. For example, turn off the grid and all errors, then turn on M2 and M2 errors and save the image. These plots help in understanding the behavior of each metal layer.

13. Iterate between the *scan rj* command and filter setting to derive a good set of filters to observe the resistor current density in the design. These filters are generally decreased in magnitude logarithmically. We can generate the plot of the current density by using the *plot rj* command.

    Look for the few wire segments that have the highest values. Create a GIF image of this plot using the *image command*. If we define the appropriate model parameters, we can also generate a plot of electromigration risk.

14. Create a plot for each layer with only the metal layer and its error layer turned on. For example, turn off the grid and all errors, and turn on M2 and M2 errors and save the image. These plots help us to identify the specific wires most likely to fail because of electromigration.

Because of the large amount of data in large designs, it is recommended to store the temporary Thunder and Lightning files in the large-size *tmp* directory of the local machine. Lightning uses the temporary files for commands. If they reside in a machine elsewhere on the network and the network is overloaded, Lightning's performance might be affected. Moving as many as files as possible to the local machine can improve its performance significantly. Use the *scan command* to iterate the filter settings in order to avoid the processing involved in plotting large volumes of the data until it is really needed. Depending on the design size and network performance, iterating significant amounts of data with the *plot command* can be slow.

It is a good practice to save the state of the analysis after finding a solution to the power grid equation, to avoid having to solve the power grid again to continue the analysis in another session.

Be sure there is enough disk space when a lot of data is processed with Lightning to avoid problems. Redrawing the grid of a large design can be time-consuming. We can use Ctrl-C to interrupt the redrawing of the power grid.

If there are different power supply voltages, which were designed in different power grids, these power grids should be analyzed separately. For each power grid, we can use the described method to do the modeling and analysis and generate the reports and plots.

Figure 5-2 shows a static power grid analysis flow application in one communication chip [53]. The "*xtc64*" is the transistor-level netlisting for the entire chip. The "*thunder.tty64*" models the peak currents for devices, and multiple (two in the flow) *thunder.tty64* commands are feasible for partitions of the entire chip in order to speed up the simulation time. The "*design.ckt*" specifies the voltage levels at the power pads.

The "*tablegen*" command generates the table of current curves for transistors according to the device sizes. The table of I-V models for the transistors is useful in the peak current simulation. The "runFEX_p_a," "runFEX_p_b," etc. extract the resistance model of the power network. Multiple resistance extraction commands (four in the flow) may be performed on the stripes of the power grid in order to speed up the computational time.

The "mergenet" command stitches the resistance models of the power network and the peak currents of devices into a complete *IR* model of the full-chip power network. Each device is modeled as a current source in the peak current, and each wire segment or via in the power network is modeled as a resistor.

Note that the flow in Figure 5-2 is applied on both the V_{dd} and V_{ss} nets, so we can estimate the worst-case *IR* drop between V_{ss}

```
#! /bin/csh -f
#This is the flow for core V_dd/V_ss nets IR static analysis
#(ocelot, 64bit, 4GB Memory, 30GB disk)
#/chip/thunder/tablegen tablegen.cmd
cd /remote/chamfs3/jonathan/simplex/viper_d_2/xt
xtc64 chip_xt.cmd
net_profile chip_cmln.net
cd /remote/chamfs4/home/qing/simplex/viper_d/thunder/static_1
thunder.tty64 run.cmd
cd /remote/chamfs4/home/qing/simplex/viper_d/thunder/static_2
thunder.tty64 run.cmd
cd /remote/chamfs4/home/qing/simplex/viper_d/thunder
itaputil combine static_1/vdd.ipeak static_2/vdd.ipeak vdd.ipeak
itaputil combine static_1/vss.ipeak static_2/vss.ipeak vss.ipeak
cd /remote/chamfs4/home/qing/simplex/viper_d/firePOWER
runFEX_p_a&
runFEX_p_b&
runFEX_p_c&
runFEX_p_d
cd /remote/cougar3/simplex/viper_d/firePOWER
mergenet _s VDD_simplex -o add /remote/chamfs4/home/qing/simplex/viper_d/firePOWER/chip_cmln_*.hdr
mergenet _s VSS_simplex -o vss /remote/chamfs4/home/qing/simplex/viper_d/firePOWER/chip_cmln_*.hdr
cd /remote/cougar3/simplex/viper_d/lightning/static_vdd
lightning.tty64 run.cmd
cd /remote/cougar3/simplex/viper_d/lightning/static_vss
lightning.tty64 run.cmd
```

Figure 5-2. Static IR analysis flow [53].

and V_{dd}. The "*lightning.tty64*" command solves the IR model for the voltage drops across the power grid.

The peak current modeling may be overestimated in the above static IR drop analysis method. Static analysis runs fast with the assumptions that all the devices are on during the chip operation, which is a worst-case assumption for the chip power consumption.

Although a dynamic analysis capability is possible by using iterative test vectors at the circuit inputs, it will take a much longer time and may not be preferred in design iterations for the purpose of power grid improvement.

When increasing the accuracy of the peak current modeling in the static current analysis, we have to provide the *current scale factor* for the chip, which is decided by us in the CAD flow to scale the estimated peak current by the tool that matches the measured or realistic current consumptions [51]. The peak current in the measurement is about 4.0 A for each of the V_{dd} or V_{ss} nets in the design.

So we can use this peak current as the baseline to match the peak current value estimated from the tool to determine the current scale factor. We found that V_{ss} and V_{dd} nets estimated from the tool have different peak current values, so the current scale factors will be different for V_{ss} and V_{dd}, since the measured current will be the same for both nets.

We do find other conclusions from the experiments on the flow in Figure 5-2. The tool does not like too many floating metals that cause the floating nodes in the power network. The disk space and memory should be very large in a large-chip power network. The turnaround time in the communication chip is about 24 hours and the automation of the running flow, as shown in Figure 5-2, helps to submit the job overnight and do the power grid improvement during the next day [53].

To reduce the extraction and analysis time, we can separate the power networks for the I/O ring and the core power network by using different V_{dd} and V_{ss} labels, so the analysis is only for the core power network in the experiments. In addition, multiple CPUs can be used to do the resistance extraction in multiple strips for the power network and the peak current estimation in multiple partitions in parallel.

The "*lightning64.tty*" command should be used in a 64-bit machine in order to run the full-chip level. The "*tablegen*" command is only needed once in a design if the circuit and process technologies are not changed. The IV table can be reused in the flow for the power grid improvement.

The first tape-out of the chip fails due to the significant *IR* drop (~0.8 V between V_{ss} and V_{dd} nets) across the chip, because the wire bonding technology is used in this chip [53]. We added a dedicated M5 for the V_{dd} and V_{ss} straps to reduce the *IR* drop.

Table 5-2 shows the voltage drops across the chip for the V_{dd} and V_{ss} networks by using separation of 40 μm and 75 μm between two adjacent V_{dd} lines or two adjacent V_{ss} lines, as well as the original design without the M5 power straps [53]. The worst-case *IR* drop calculation is the sum of the voltage drops in the V_{dd} and V_{ss} nets.

Table 5-3 shows the current scaling factors in the simulation for the V_{dd} and V_{ss} nets in order to match the currents simulated by the tool with the peak current assumption (4.0 A), based on the measurement from the original tape-out chip in the same process technology and not significantly changed circuits.

Table 5-2. Simulation results

	V_{dd} IR Drop	V_{ss} IR Drop	Worst-Case $(V_{dd} + V_{ss})$ IR Drop
Original Chip: No M5 Power Straps	0.356 V	0.434 V	0.79 V
Additional M5 Power Straps: 75 μm Separation	0.117 V	0.146 V	0.26 V
Additional M5 Power Straps: 40 μm Separation	0.130 V	0.124 V	0.25 V

Table 5-3. Current scaling factors in simulation [53]

	Simulated Current	Measured Current	Current Scaling Factor
V_{dd}	4.22 e + 03 A	4 A	0.00095
V_{ss}	5.25 e + 03 A	4 A	0.00076

About 67% IR drop reduction is observed by adding the power straps on the M5 layer for this chip, and the 0.25 V IR drop is within the required supply voltage ranges (nominal voltage: 2.5 V) for the correct device timing, which is about 10% of the nominal voltage.

The old design has an IR drop ($V_{dd} - V_{ss}$) of about 30% of the nominal voltage, which is one reason that the chip fails. Figure 5-3 shows the voltage plots of the power grids for the old design without M5 power straps. We can observe significantly low voltage at the center area of the chip.

5.3 ADVANCED STATIC ANALYSIS

Activity-based analysis is another approach to static analysis that better resolves the distribution of currents on the power grid [51]. The activity-based approach assumes that you have a mechanism, such as a Verilog simulator, to compute and report the relative activity of the nets in the design.

These relative activities can be used in conjunction with net capacitances to estimate the average current load of each gate in the design. This form of analysis will provide a more realistic power current than the Ipeak estimation approach based on the saturation currents.

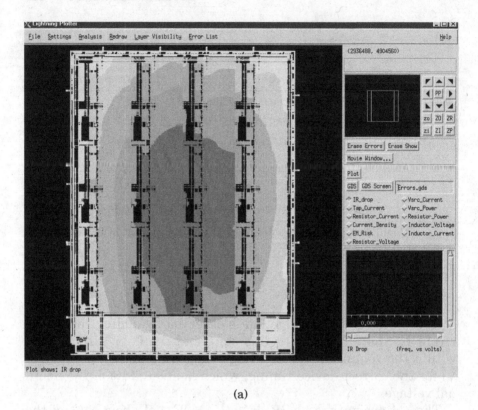

(a)

Figure 5-3. Voltage plots of power distribution networks [53]. (a) V_{dd}.

As input, the activity-based analysis uses a file containing the activity levels of nets in the design. This file is optional but recommended in VoltageStorm™ [51]. In addition, activity-based analysis has three other important input parameters: (1) the clock cycle time of the chip for which the activity values are defined, (2) The value of V_{dd}, and (3) the default activity to use for gates whose activity is not specified.

The average current for each gate is computed using the loading of the gate, V_{dd}, the cycle time, and the activity of the gate. The average current consumed by a gate is derived by the following equation:

$$I_{\text{avg}} = A \cdot C_{\text{gate}} \cdot V_{dd} \cdot F \qquad (5\text{-}1)$$

where A is the activity ratio of the gate, C_{gate} the total capacitance of the nets including the wires and gates, F the clock fre-

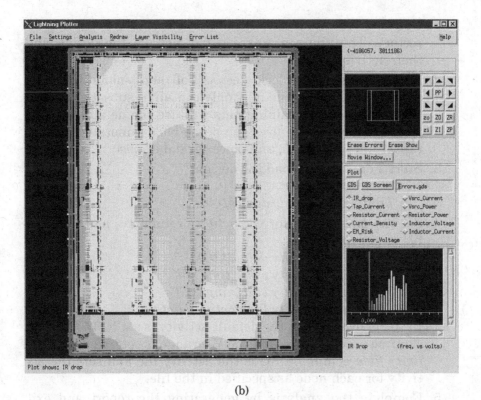

(b)

Figure 5-3 *(continued)*. (b) V_{ss}.

quency of the chip, and V_{dd} the supply voltage. This equation for
the average current is derived by considering the charge, Q, required to charge the outputs of the gate in a clock cycle interval
$(1/AF)$.

This derivation of average current is not a function of transistor
sizes. If your design has multiple clocks, select one clock to be the
reference for the activity analysis, and scale the gates associated
with other clocks accordingly.

For example, if CLK1 has a period of 10 ns and CLK2 has a period of 15 ns, and CLK1 is to be the reference for activity-based
analysis, scale the activity of gates in the CLK2 domain by 0.666.
On the other hand, if we have the actual toggle numbers of all
nets, use CLK1 as the reference to divide the net toggle counts to
derive the activity values.

Here are steps to run the activity-based static power grid
analysis using the VoltageStorm™ tool [51].

1. Obtain a file called *activity.list* and put it in the running directory. This file contains a list of nets in the design and their activity levels.

2. For the nets in the clock trees, all of them should have the activity of 1.0. This file may only contain a subset of nets in the design, and remaining nets will use the default activity factor. This analysis uses the name back-annotated netlist, capacitance database, and power grid database.

3. Start Thunder and load the design:
   Shell >> *thunder*
   Thunder > *load design.ckt*

4. Set the parameters for the *activity* command, and load the file of activity values.
   Thunder > *activity default 0.03*
   Thunder > *activity cycle_time 5ns*
   Thunder > *activity vdd_range 3.3*
   Thunder > *activity filen activity.list*

In the above setting, the default activity is set to 3%. The cycle time is set to 5 ns, and the V_{dd} range is set to 3.3. The *activity filen* command reads the activity file and sets the activity for each node as specified in the file.

5. Complete the analysis by generating the report and exit Thunder. The *activity report* command computes the tap currents on the basis of these activities, based on Equation (5-1), and writes them in the *VDD.iavg* file:
   Thunder > *activity report VDD*
   Thunder > *quit*

6. We now proceed to the power grid analysis and change to the Lightning directory. Copy the files from the static directory for defining the voltage sources (*vsrc.cmd*) and defining filters (*filters.cmd*) into the Lightning directory. Start Lightning, load the grid, define the voltage sources, and load the currents, which have just been computed:
   Shell >> *lightning*
   Lightning > *load design_VDD.mhdr*
   Lightning > *run vsrc.cmd*
   Lightning > *run filters.cmd*
   Lightning > *iload $thunder_dir/VDD.iavg*

The above commands load the necessary information for a power grid solution. We can view the tap currents computed from the activity information in the chip by the following command:

Lightning > *plot tc*

7. Now solve the grid, and then plot the *IR* drop (*plot ir command*) and the resistor current (*plot rc* command):

Lightning > *solve*
Lightning > *plot ir*
Lightning > *plot rc*
Lightning > *quit*

The third method for performing static power grid analysis is to use dynamic vectors to exercise the design when computing the average currents in the transistors connected to the power grid. This method will compute the average currents for transistors on the basis of the specific vector set. The vector set must be sufficiently representative of the design usage to achieve accurate average currents.

VoltageStorm™ allows three ways to specify input vectors: (1) define SPICE-like voltage sources in the netlist, (2) use a Thunder-specific vector file to describe waveforms, and (3) use the VCD file.

The SPICE-like voltage sources that Thunder supports are DC, pulse, and piecewise linear (PWL). The following steps use the SPICE-like voltage sources to drive only the clock to illustrate how to perform the vector-based netlist analysis.

1. Prepare a *design_input_sources.inc* file in the Thunder working directory. This *design_input_sources.inc* file includes the pulse waveforms for CLK and CLKN with a period of 5 ns. Only the clocks are used in this example to illustrate the vector-based static power grid analysis.

2. Start Thunder and load the circuit:

Shell >> *thunder*
Thunder > *load design.ckt*

Simulate for two clock cycles in the DC state to initialize the system properly. After computing an initial state, we can save it to reuse later:

Thunder > *s 30*
Thunder > *save ic state.ic*

The *s* command performs the DC solution and performs the circuit for 30 ns, which is in two clock cycles. The *save ic* command saves the voltages of the circuit in the format of .IC cards. This file is used in the dynamic analysis to avoid the computation of a DC solution.

3. The following commands compute the currents for devices connected to the V_{DD} source:

> Thunder > *devi tally VDD*
> Thunder > *devi tran VDD*

The *devi tally* command instructs Thunder to begin tracking the minimum, maximum, and average current for the V_{dd} voltage source when you next perform the simulation. The *devi tran* command instructs Thunder to create a Thunder.tran output file that provides the transient waveform for the currents of the voltage source V_{DD}.

Simulate for another 10 ns in two clock cycles, report the tailed currents for V_{dd}, and exit Thunder:

> Thunder > *s 10*
> Thunder > *pwrnet report VDD*
> Thunder > *devi report*
> Thunder > *quit*

The pwrnet report command instructs Thunder to write the currents reported so far into the VDD.avg, VDD.max, and VDD.rms files. In this case, the three files are all in ASCII format [51].

The *devi report* command instructs Thunder to report the minimum, maximum, average, and RMS currents of the V_{DD} voltage source. According to the SPICE convention, the current of a device entering the device at the terminal is positive, so normal current flow into the V_{dd} voltage source is negative. Therefore, the reported minimum value is the peak absolute current generated by the V_{dd} source, and the average current should be negative.

4. The power grid analysis portion of the flow is much like the analysis performed in the activity-based static analysis, except that the currents input file is different:

> Shell >> *cd $lightning_work_dir*
> Shell >> *lightning*

Load the design and voltage sources command files as follows:

> Lightning > *load design_VDD.mhdr*
> Lightning > *run vsrc.cmd*

5. Load the current input file from the vector-based simulation result, and then perform the power grid analysis as follows:
   Lightning > *iload VDD.iavg*
   Lightning > *solve*

6. The above power analysis performs the static analysis by averaging the currents over clock cycles. The *IR* drop result is based on the average currents over clock cycles. On the other hand, we can use the VDD.max file, which tracks the peak current of each transistor on the power grid. We can analyze the power grid again using the VDD.max file, after clearing the early current inputs by the iclear command, as follows:
   Lightning > *iclear*
   Lightning > *iload VDD.max*
   Lightning > *solve*
   Lightning > *quit*

When we use the VDD.avg file, for example, the *IR* drop in this case goes down as far as 3.265 V [51]. However, when we use the VDD.max file, the *IR* drop goes down to 2.772 V [51], but this number may be an overestimation of the peak *IR* drop in the power grid, because it models all the transistors turned on to their maximum currents at the same time.

The actual peak *IR* drop is somewhere between that reported using the VDD.max file and that reported using the VDD.avg file. Use the VDD.max file only on the small blocks to perform an easy pass and fail screening of the block power grid.

If we want to apply peak currents to large designs to which many vectors have been applied, we will see an unrealistic measure of the *IR* drop. We can use it on small blocks in which many gates could potentially switch at the same time.

5.4 DYNAMIC ANALYSIS

The dynamic analysis method is claimed to provide more precise insight into the behavior of the power grid [51]. The static analysis averages the tap currents to look at the long-term average behavior of the power grid. Dynamic analysis keeps the time distribution of currents in place so you can see the voltage and current waveforms in a more numerically precise way.

Therefore, the dynamic analysis will provide better insight into the magnitude of the *IR* drop. The dynamic analysis capability provided by the VoltageStorm™ tool is claimed to have the following goals [51].

It helps the designer to find the weak spots in the power grid by predicting a worst-case test vector for the *IR* drop from the test vectors that we have. It is usually hard to find the worst-case *IR* drop test vector because it is a function of the physical implementation of the design, not the logic implementation.

The dynamic analysis enables us to analyze the specific test vectors on the design. This capability is most useful when we know some specific test vectors that we must analyze in great depth to obtain the exact magnitude of *IR* drop in the power grid.

It is critical to select the proper step size for the power grid analysis. This step size in the power grid analysis is different from the simulation time step in the netlist simulation. The netlist simulation uses internal time step control to keep the simulation accurate. The power grid analysis step size reflects how often the tap currents pass from the netlist analysis to the power grid analysis.

As described before, the VoltageStorm™ tool uses vector compression technology to speed up the dynamic analysis for a large design. Obtaining usable results from the vector compression requires us to set the parameters for the compression carefully.

We can set three parameters as follows:

1. *Method:* determining whether to compress using peak or averaging across multiple vectors
2. *Period:* the time period over which we want compression to be applied
3. *Intervals:* the number of time steps that we want in each period

In general, we use the clock period as the period of compression because we want to gain a better insight into the operation of the chip over a clock cycle. Most circuit activities occur near the edges of the clock, so we want to see if *IR* drop problems occur because of the clock itself or as the result of logic switching after the clock.

The second question to address is the number of intervals to apply. Once again, the starting point is based on the delay of a typical gate. The number of intervals or timing buckets is the period divided by the gate delay. It is always better to use more intervals to solve the power grid if the computational time is tolerated for a large power grid.

The third question is the selection of the method to apply in the vector compression: peak or averaging. If the design is small and could possibly have more simultaneous activity than is represented by the vectors, we want to use peak compression. Increasing the intervals is necessary for large designs because the finer time stepping reduces the overestimation of *IR* drop resulting from bucketing activity at the same time, when in reality it occurs at different times.

If we use peak compression, we need to use the average of peak currents in the power grid analysis. Do not use peak compression if the design contains exclusive logic, in which at most one in n components could ever operate at once.

The average-to-peak currents form a good data set if the time steps are small enough that the peak current in a time step does not highly overestimate the average current in the time step.

The dynamic analysis is the next step in complexity beyond the vector-driven static analysis. The dynamic analysis uses vectors as input to the netlist analysis performed by Thunder. It generates the dynamic current data to feed into the power grid analysis performed by Lightning. Here are the steps for one example to perform the dynamic analysis, based on the VoltageStorm™ tool [51]:

1. It needs a VCD-format input file (*inputs.vcd*), and then we start Thunder in the Thunder directory as follows:
   Shell >> *thunder*
   Thunder > *load design.ckt*
2. Enter the following command to tell Thunder to use the initial conditions specified in the file when we begin the simulation.
   Thunder > *use ic state.ic*
3. Perform the vector compression using the peak function across the vectors and compute 20 time steps in the 5 ns period:
   Thunder > *pwrnet tallyint method=000 intervals=20 period=5ns VDD*
   In this case, because simulation runs for 10 ns, two vectors are compressed into one and each power grid analysis time step is 250 ps wide. Tally the current from the V_{dd}

source, create a Thunder.tran output file, submit a VCD-format file, and report tallied currents for V_{dd}:

> Thunder > *devi tally VDD*
> Thunder > *devi tran VDD*
> Thunder > *vcd inputs.vcd*
> Thunder > *devi report*
> Thunder > *quit*

The above simulation generates several files. The *pwrnet tallyint* command creates three files: VDD.ptimax, VDD.ptiavg, and VDD.ptirms. They correspond to the peak, average, and RMS currents, respectively, for each interval of analysis. The three files correspond to the peak-to-peak, peak-to-average, and peak-to-RMS currents. The *devi tran* command creates the Thunder.tran file containing the transient waveform for the V_{dd} voltage source current.

4. After the completion of the netlist analysis, run the following commands:

> Shell >> *itaputil summary VDD.ptiavg*
> Shell >> *itaptuil s VDD.ptimax*

The *itaputil summary VDD.ptiavg* command generates a summary of the current data in the VDD.ptiavg file. There are 20 intervals of the data in this example, and the report shows the minimum, maximum, and average currents over all transistors connected to V_{dd}, as well as the total V_{dd} current in the interval.

5. Given the above dynamic current files, we are ready to proceed to the power grid analysis. Dynamic power grid analysis is similar to static analysis. Dynamic analysis performs a series of power grid matrix solutions, one for each time step. Currents are updated for each time step and capacitance models are updated. The resulting states for each solve are saved automatically.

   Change to the Lightning working directory, and start the Lightning command from there:

> Shell >> *lightning*
> Lightning > *load design_VDD.mhdr*
> Lightning > *run vsrc.cmd*
> Lightning > *run filters.cmd*

The above commands are the same as the static analysis to load the design database of the power grid, the V_{dd} source locations, and the filters.

6. The following command specifies the current file to apply and initiates the dynamic analysis:
   Lightning > *tran VDD.ptiavg*

The *CurrentScaleFactor* environment variable is not set in this dynamic analysis. If we set the value for *CurrentScaleFactor* environment variable, it would scale the currents appropriately. We could set it to overestimate the power currents to compensate for the averaging of peak currents resulting from taking averages either within each time step or across vectors in the compression.

We can also set up the filters to generate the plots and reports during the dynamic analysis. The VoltageStorm™ tool also provides a movie of the behavior of the power grid over the time intervals of the analysis. We can examine the plots and reports of the individual states as in the static analysis.

The *tran* command computes the state of the power grid after each time step during the analysis. These states are saved in a set of files sequentially numbered beginning with Lightning. tran_int0. We can load each of these states individually by using the *loadstate* command and generate plots and reports of these individual states.

The most useful state created during the dynamic analysis is the *Lightning.worstcase* state. It contains the worst-case voltages over the dynamic analysis for each subnode in the power grid, so you can examine a single file to determine the worst *IR* drop occurring in the dynamic analysis.

5.5 LAYOUT EXPLORATION

VoltageStorm™'s power grid exploration capability enables the designer to optimize the power grid or correct a problem inside the database [51]. We can experiment with power grid changes, such as adding or changing vias, voltage sources, or resistors, and perform the power grid analysis to show the effects of these changes in the power grid.

The power grid layout changes are easier to complete if no signal routing is completed. The PGS exploration is used once we have placed all cells and transistors and have a complete physical power network.

We do not have to wait until we have completed the signal routing. Because we can easily explore the effects of changes to a power grid network in the PGS exploration framework, we can determine if we have overdesigned the power grid. Although some overdesign is necessary, significant overdesign decreases the available signal routing area and wastes the die area.

PGS exploration lets us rapidly understand the consequences of reducing or increasing power route widths, so we can adjust the power grid design to the power grid requirements.

Once we load the power grid and tap currents into Voltage-Storm™, we can use PGS exploration [51] to modify the power grid resistor network as desired. When we want to understand the effects of the changes, we simply perform a power grid solution. We can continue to repeat the modification and solution steps until the power grid is clean.

When we are satisfied that the power grid design is acceptable, we write out the change report from VoltageStorm™ and use it as the guide for implementing the changes in the layout. A *change report* contains a summarized list of the changes made to the power grid. The changes are made to the resistors within Voltage-Storm™.

The change report is written in the layout format, which contains the width, length, layer, and coordinate information that enables a layout designer to easily implement the required layout changes to the power grid. In order to avoid redundant layout changes when a resistor is modified more than once, the change report lists only the final modifications.

PGS exploration includes the following commands:

1. *addres:* adds a resistor to the power grid
2. *changeres:* modifies an existing power grid resistor
3. *addvia:* adds a new via to the power grid
4. *addvsrc:* adds a voltage source to the power grid
5. *show:* displays the selected nodes or elements
6. *unselect:* deselects the signal or multiple nodes or elements
7. *write:* writes out the change report

Most of the above commands allow us to modify the power grid network.

Typically, we can select the object to modify and then execute the change command. We can select objects interactively by using the middle mouse button and drawing a selection box over the area to be selected. To be selected, an object must be completely inside the selection box. We can use the *select command* to select the objects. We can select resistors interactively by clicking on them with the middle mouse button [51]. The following are the steps for one example using PGS exploration.

1. Move to the directory containing the V_{dd} power grid database, and start Lightning from there:
   Shell >> *lightning*

2. Load the power grid database into Lightning as follows:
   Lightning > *load design_VDD.mhdr*
   Add the voltage sources to the power grid using the specified command file as follows:
   Lightning > *run vsrc.cmd*
   Load the tap currents into Lightning as follows.
   Lightning > *iload VDD.ipeak*

3. We can solve the power grid equation by using the following command:
   Lightning > *solve*
   Use the auto-filtering to define the filter ranges for the *IR* drop, and display the *IR* drop in the Lightning plotter window as follows:
   Lightning > *filter ir auto*
   Lightning > *plot ir*

4. Use autofiltering to define the filter ranges for the *IR* drop, and display the *IR* drop in the Lightning plotter window as follows:
   Lightning > *filter ir auto*
   Lightning > *plot ir*
   In this example, the large *IR* drop could be observed in the left side of the central control section. It occurs because the power routing to this block comes only from the right side. In order to fix the problem for this larger *IR* drop, a resistor could be added on the M1 layer to connect the upper row of the control block.

5. First zoom in on the upper left corner of the central block by using the left mouse button to draw the box from the lower left to the upper right around the area. Select a node by clicking on it, and then add a resistor between the selected nodes using the following command:
   Lightning > *addres selected 1000*

This command adds a resistor with a width of one micron (1000 units) between two selected nodes. Notice that the resistance of this resistor is automatically calculated from the process technology information.

Repeat the *addres* command to add additional resistors to the adjacent nodes in the power grid.

6. Solve the circuit again and replot the *IR* drop. The reduced *IR* drop can be seen to be due to the added metal lines on the power grid:

    Lightning > *solve*
    Lightning > *plot ir*

7. Now verify that the current density limits have not been exceeded after modifying the resistance in the power grid. The autofiltering can be used to set the filters for the current density and plot the current density errors:

    Lightning > *filter rj auto*
    Lightning > *plot rj*

    The red color in the upper right corner indicates that the current density has exceeded the limits. Change the width of the resistor in the upper right corner by a factor of 5.0, deselect all, solve the circuit, and plot the current density errors as follows.

    First select a resistor of the red color by clicking on it with the middle mouse button. Then we perform the following commands:

    Lightning > *changeres selected 5.0*
    Lightning > *unselect all*
    Lightning > *solve*
    Lightning > *plot rj*

    We can see that changing the width of the resistor by a factor of 5 made a significant improvement in the current density. Then we replot the *IR*, since the resistance of the power grid has been changed:

    Lightning > *plot ir*

8. Next, we will generate the reports. The following command will generate a report with a list of all resistors that have been changed:

    Lightning > *changeres report*

    The following command will generate a report with a list of all added resistors:

    Lightning > *addres report*

The following command writes out the change report to guide the layout changes:

Lightning > *write gridchanges design_layout.eco*

The above command creates a file named *design_layout.eco* that contains all the commands that make changes to the power grid. Then we need to quit Thunder to finish the ECO changes in the power grid:

Lightning > *quit*

5.6 SUMMARY

With the complexity of the power grid and reduced power supply voltages in modern VLSI chips, CAD tools are necessary to assist designers in finding failures or weak spots in power network designs. This chapter discusses the most popular tool, VoltageStorm™ from Cadence, with modeling and analysis capability, and explains how to use this CAD tool to aid in *IR* drop analysis and improvement.

The tool provides the following capabilities: (1) modeling of the power network in the resistance network, (2) modeling the transistor switching current in the tap current, and (3) solving the power network model in the linear circuit.

The tool also provides the capability to help designers locate and fix errors in the power grid layout. For example, PGS exploration is one example that uses the internal power grid analysis database to fix the power grid and output a list of changes needed with zero violations for the power grid. Layout designers can use up these changes, as necessary, to fix the power grid design.

6

MICROPROCESSOR DESIGN EXAMPLES

Microprocessor chips usually consume a lot of power and therefore have the highest requirements for power network distribution performance. This chapter contains seven sections. Section 6.1 describes the Intel IA-32 Pentium-III chip [66]. Section 6.2 describes the Sun UltraSPARC chip [67]. Section 6.3 describes the Hitachi SuperH microprocessor chip [68]. Section 6.4 describes the IBM S/390 microprocessor [69]. Section 6.5 describes the Sun SPARC 64b microprocessor [70]. Section 6.6 describes the Intel IA-64 microprocessor [71]. Section 6.7 summarizes this chapter.

6.1 INTEL IA-32 PENTIUM-III

The Intel IA-32 microprocessor is implemented in a five-layer metal 0.25 μm CMOS process technology [66]. Table 6-1 shows the process technology parameters and operating voltage range for this processor. The 10.1 × 12.1 mm² die contains 9.5 million transistors. The functional unit-level local interconnects are routed using lower metal layers with higher density, whereas the global interconnects have been routed in the upper layers, M4 and M5, which have lower metal resistance. The top metal layer (M5) supports all of the C4 bumps.

Alternative power and ground grids are implemented in M5 and M4 for global power distribution. Spacing and width of these metals were selected such that inductive effects are minimized

Power Distribution Network Design for VLSI, by Qing K. Zhu
ISBN 0-471-65720-4 © 2004 John Wiley & Sons, Inc.

Table 6-1. 0.25 μm CMOS process technology [66].

Gate oxide thickness	40 A
Gate length	0.20 μm
M1 pitch	0.61 μm
M2 pitch	0.88 μm
M3 pitch	0.88 μm
M4 pitch	1.73 μm
M5 pitch	2.43 μm
Operating voltage	1.4–2.2 V

and both AC and DC drops are reduced. For the local metal layers, a tree-based distribution was chosen, with custom width selection for the trunks and branches according to the area current drain requirements. The global power grids and associated local tree structures are shown in Figure 6-1 [66].

It is difficult to optimize the power distribution using a single C4 bump pitch for both the I/O and the core due to their different requirements. In the core, the optimization is primarily driven by the potential for power collapse but constrained by the effective routing channel space available for global signals. However, in the I/O area, power collapse, minimization of the interconnect length to a C4 bump, and package-level routability are some of the additional constraints.

A 252 μm bump pitch for the core and 235 μm bump pitch for the I/Os were chosen [66]. The overlap region between the core and I/O area is strapped with custom power grids. In the I/O ring design, special attention was paid to the placement of signals and power/ground bumps and their ratio, such that loop inductance is minimized while maintaining the continuous return paths for I/O signals.

The processor is packaged in a six-layer organic land grid array (OLGA) package. Dedicated power and ground planes are used to minimize the package-level power distribution and the noise due to package-level power distribution. Power distribution was designed with two different V_{cc} supplies to enable lower-power applications.

The core power supply voltage level can be dropped significantly while maintaining the I/Os and other special analog circuits with a different supply. All of the special circuits within the core were verified at a 1.1 V supply voltage to enable this voltage scalability.

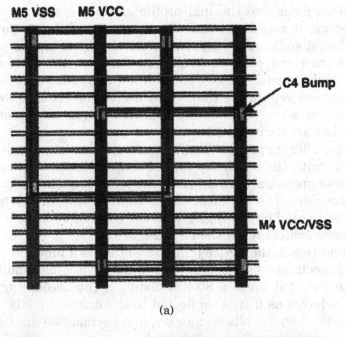

(a)

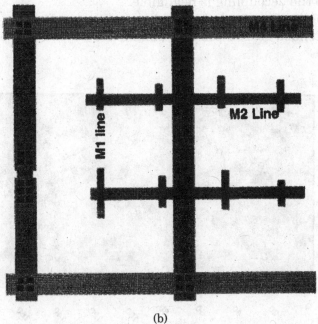

(b)

Figure 6-1. (a) Global power grid (M4 and M5) and (b) local power trees for the Intel IA-32 Pentium-III chip [66].

From a measured thermal profile of the previous Intel micro-processors, it was found that the voltage level due to power collapse is not sufficiently uniform across the die to hit the projected goal of the required clock frequency. A power distribution model was developed such that we could study the power collapse in different areas separately [66]. Knowing the worst-case switching activity per area, the coupling capacitor requirements on a per-area basis are derived [66].

Various design profiles for the process technology are derived to come up with the proximity roll-off characteristics. When designing these optimizations, a broad range of frequency components were considered in the modeling to capture several spectral components created by the high-frequency edge rates associated with transistor switching, as shown in Figure 6-2.

In Figure 6-2, the device junction voltage is a function of decoupling capacitance distance for worst-case switching conditions. It is observed that up to a 80 μm distance, the decoupling capacitance behaves as if it is connected to the driver directly. Beyond 80 μm, the impact rolls off quickly, and beyond 200 μm its contribution to the decoupling is negligible.

For the best case when neighbors are not switching, the roll-off is extended to 100 μm but diminishes beyond 200 μm. With

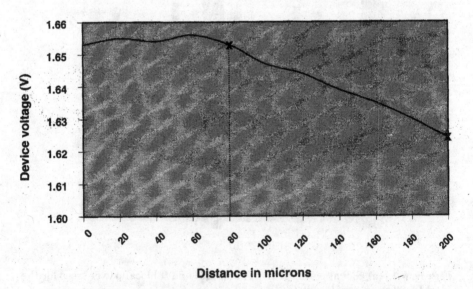

Figure 6-2. Decoupling capacitance effects on device voltage [66].

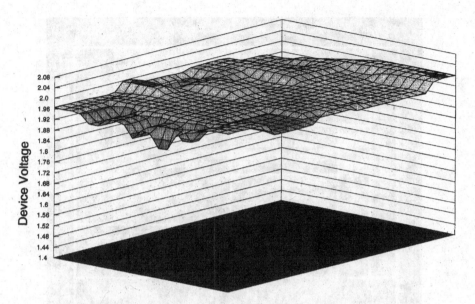

Figure 6-3. Device voltages across microprocessor chip [66].

better placement guidelines from the decoupling capacitors, as shown in Figure 6-2, a more uniform power collapse is achieved in spite of nonuniform current drain at various parts of the die. Figure 6-3 shows the power fluctuations of the chip at various points.

6.2 SUN ULTRASPARC

A 1.1 GHz 64-bit UltraSPARC microprocessor has been described in [67]. It is built on a 0.13 μm 7LM Cu CMOS process from Texas Instruments Inc. The nominal channel length for the gate is 65 nm and interconnects use the low-k dielectric with dielectric constant 3.6. The power consumption is 53 W at 1.1 GHz and 1.3 V supply voltage [67].

The die size is 178.5 mm^2. The total transistor count of this chip is 87.5 million, of which 63 million are in the SRAM cells. The chip package is a 950 pin flip-chip micropin grid array (μPGA). The signal-to-power pin ratio is 5:1 in the I/O distributions.

Figure 6-4 is a die micrograph showing the floor plan of the main functional blocks in this chip. The L2 Cache or SRAM cells

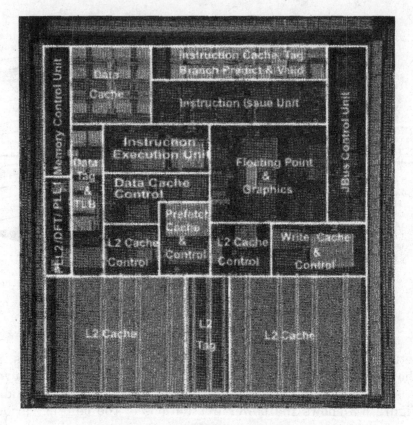

Figure 6-4. Sun UltraSPARC die micrograph [67].

are located at the bottom of the chip. The control and execution units are located in the middle of the chip.

The instruction catch and decoder blocks are located at the top of the chip. The clock is distributed from the PLL output up to the flip-flops through a balanced tree network. All the inputs of flip-flops and clock buffers are connected through a clock grid network to minimize clock skew.

The main power network uses a grid in M5/M6 and M7 (three metal layers). There are 2065 solder bumps, of which 1251 are used for V_{dd} and V_{ss}. These bumps are area-distributed over the chip area by the flip-chip technology. The I/O contains 800 solder bumps, 470 of which are signal bumps, whereas 330 are used for power and ground. The bumps in the core area and in the channel regions are placed away from the active circuitry to prevent soft errors due to alpha particles released from the bumps.

6.3 HITACHI SuperH™ MICROPROCESSOR

A 200 MHz 0.2 μm Hitachi SuperH™ microprocessor has been described in [68]. The microprocessor is fabricated with a 0.2 μm, five-metal, dual-oxide-thickness, triple-well CMOS technology. It has five levels of metal (M1, M2, M3, M4, and M5). The last two metals are thicker than the first three and the top metals are usually used for the global power distribution. The dual-tox structure enables the use of MOS transistors with two distinct tox and threshold voltages for both pMOS and nMOS devices.

Table 6-2 shows the process and device parameters used for this processor. Thin-tox, low-threshold voltage devices are provided for the 1.8 V internal circuitry, and thick-tox, high-threshold voltage devices are used for the 3.3 V circuitry, such as the I/O circuitry.

Figure 6-5 shows the pMOS and nMOS device layers and structures. Substrate biases, denoted as vbp and vbn in Figure 6-5, for the thin-tox, low-threshold voltage devices are controlled through the switched substrate impedance scheme. The substrate biases for the thick-tox, high-threshold voltage are connected to their local source terminals as in the conventional CMOS devices.

In the standby mode, the substrates for the pMOS and nMOS devices are biased to 3.3 and −1.5 V, respectively, to increase the threshold voltages of the MOS transistors and lower the subthreshold leakage current. The substrates for the pMOS and nMOS devices are biased to 1.8 V and 0 V, respectively, in the active mode to maintain high-speed operation.

The high-speed switching of MOS transistors induces significant power supply noise and local substrate noise. This noise makes it difficult to bias the substrate of all MOS transistors uniformly. In

Table 6-2. Process and device parameters for the Hitachi SuperH™ CPU [68]

Technology	0.2 μm, P-sub, triple-well CMOS
Gate channel length (Lg)	0.2 μm (1.8 V device) and 0.35 μm (3.3 V device)
Gate oxide thickness (tox)	4.5 nm (1.8 V device) and 8 nm (3.3 V device)
Threshold voltage (Vth)	0.15 V (1.8 V device) and 0.45 V (3.3 V device)
Metal layers	Metal 1–3 (0.88 μm pitch) and Metal 4–5 (1.76 μm pitch)
Area	6.84 × 6.84 mm²
Transistor count	3.3 M

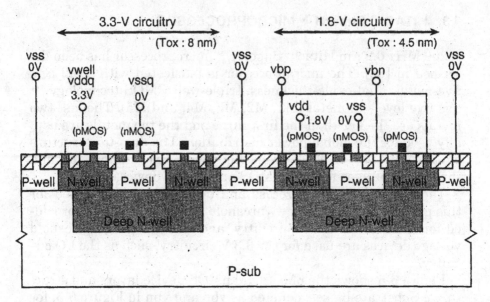

Figure 6-5. pMOS and nMOS device structures [68].

the active mode, the fluctuation in the substrate bias causes significant threshold voltage variation and lowers the operating speed.

The peak overshoot of the substrate noises can be reduced by lowering the supply voltage or increasing the source and substrate diffusion capacitances. The decap time of the noise depends on the substrate impedance. A long decap time that exceeds the cycle time causes the substrate noise to accumulate.

To reduce the substrate impedance and achieve substrate biasing, the switched substrate impedance scheme has been developed. This scheme switches the substrate impedance, as well as the substrate bias, according to the operation mode. Figure 6-6 shows the switched impedance scheme for this microprocessor.

A standby controller and a v_{bb} controller (VBC controller) control the voltage of the substrates, denoted as v_{bp} for the pMOS substrate and v_{bn} for the nMOS substrate. In the standby mode, these are driven with a high-voltage, high-output impedance driver in the VBC macro. In the active mode, the substrates are driven with about 10000 switch cells over the chip [68].

Each switch cell consists of two thick-tox and high-threshold voltage MOS transistors. One transistor with a gate signal c_{bp} is connected to v_{bp} and a_{dd}. Another with a gate signal c_{bn} is connected to v_{bn} and v_{ss}. These transistors reduce the substrate imped-

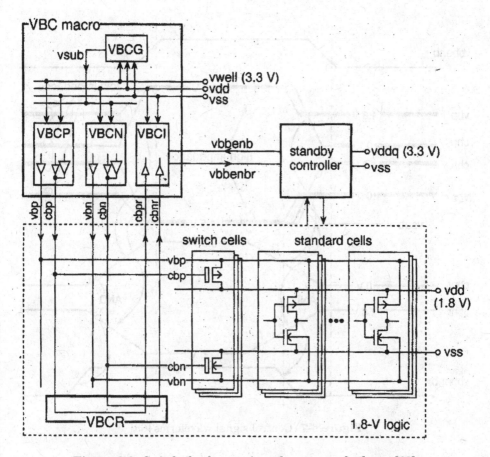

Figure 6-6. Switched substrate impedance control scheme [68].

ance; in other words, they keep the substrate biases of the MOS transistors equal to their local power supplies.

Therefore, even if the local power supply drops due to a power line pump or simultaneous switching noise, the substrate bias is quickly recovered. The VBC macro consists of four circuits—VBCP, VBCN, VBCI, and VBCG—and is fed by supply voltages a_{dd} (normally 1.8 V) and v_{well} (3.3 V). VBCG generates v_{sub} voltage, which is a negative voltage used as the third voltage source in the VBC macro. The v_{sub} voltage is equal to $a_{dd} - v_{well} = 1.8$ V $-$ 3.3 V $= -1.5$ V.

Figure 6-7 shows the waveforms of a complete transition from active mode to standby mode. When the microprocessor goes from the active to the standby mode, the standby controller stops all

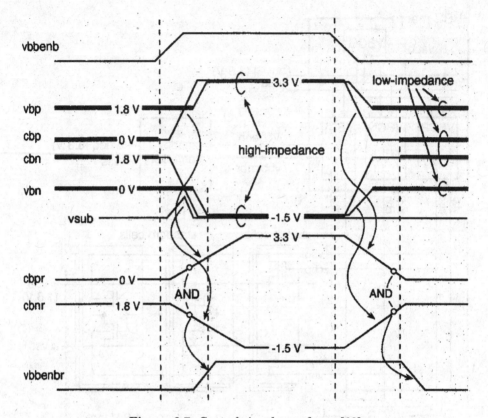

Figure 6-7. Control signal waveforms [68].

1.8 V logic circuits. After that, it issues a *vbbenb* signal. Then the VBC macro drives c_{bp} up to v_{well} (3.3 V) and c_{bn} down to v_{sub} (−1.5 V). These signals turn off all switch cells. The VBC macro also drives c_{bp} to 3.3 V and v_{bn} to −1.5 V. This mode transition takes about 50 μs.

Figure 6-8 shows the layout of a standard cell and a switch cell for random logic circuitry. Both cells have the same height. In a conventional CMOS cell, the substrate biasing lines, v_{bp} and v_{bn}, are connected to the power lines (a_{dd} and v_{ss}) locally. In the new scheme, these lines are interconnected separately to bias the substrate.

The substrate bias lines v_{bp} and v_{bn} are interconnected by M1 and are parallel to the power lines a_{dd} and v_{ss}. The switch cell has additional vertical power lines a_{dd} and v_{ss} interconnected by M2. Furthermore, between a_{dd} and v_{ss}, there are four metal lines: two

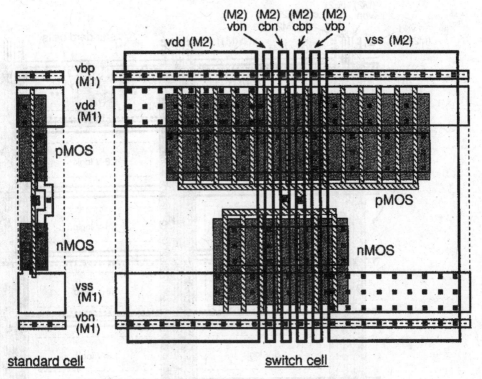

Figure 6-8. Standard cell and switch cell layouts [68].

are the substrate biasing lines v_{bp} and v_{bn} and the other two are the gate lines c_{bp} and c_{bn}.

In order to reduce the chip area overhead, the design uses identical heights for each cell compared to the conventional CMOS cell, as shown in Figure 6-8 [68]. The width of the power lines to M1 is reduced to about 77% that of the conventional CMOS cell. This increases the impedance of the power lines.

To reduce the impedance, the power lines are routed in a fine mesh structure. Figure 6-9 shows the metal routing of v_{bp}, v_{bn}, c_{bp}, c_{bn}, and power lines. The switch cells are placed in rows, and the distance between two switch cells is about 200 μm. The thicker metal levels of M4 and M5 also form a coarse power line mesh that reduces the impedance of the power lines. The chip area overhead of the switch cells is less than 2% because the switch cells are placed under the power lines in M2, as shown in Figure 6-9.

The data flow in the data path is designed so as to be parallel to the power lines and p- or n-wells. This layout will reduce the

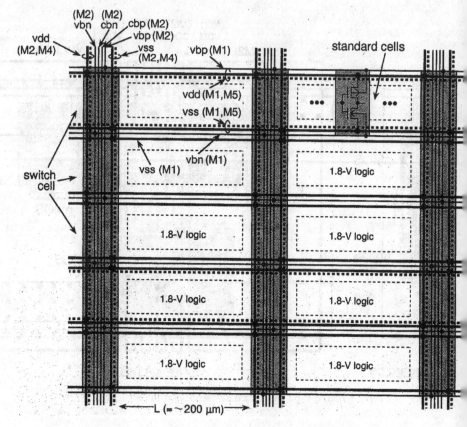

Figure 6-9. Power grid structure for microprocessor [68].

number of logic cells operating on the same well simultaneously. It also reduces the injected noise. The substrate biases of 3.3 V for pMOS and −1.5 V for nMOS decrease the subthreshold leakage current during the standby mode by about 1.5 orders of magnitude. However, a larger body effect degrades the circuit performance by elevating the threshold voltage in series-connected MOS transistors or pass transistors.

6.4 IBM S/390 MICROPROCESSOR

A microprocessor implementing IBM S/390 architecture operates at frequencies up to 411 MHz (2.43 ns). The chip is fabricated in a 0.2 μm L_{eff} CMOS technology with five layers of metal and tung

Table 6-3. S/390 microprocessor technology parameters and chip characteristics [69].

L_{eff}	0.2 μm
Gate Oxide	5.5 nm
M1 Pitch	1.2 μm
M2 Pitch	1.8 μm
M3 Pitch	1.8 μm
M4 Pitch	1.8 μm
M5 Pitch	4.8 μm
Power supply	2.5 V
Transistor count	Logic (3.8 million)
	Array (4.0 million)
Die size	17.35 mm × 17.3 mm
Power	37 W @ 2.5 V 300 MHz
Maximum frequency	411 MHz
Area C4	1600
Off-chip signal I/O	448
On-chip decoupling capacitance	102 nF

sten local interconnects. The chip size is 17.35 mm × 17.3 mm with about 7.8 million transistors. The power supply is 2.5 V and measured power dissipation at 300 MHz is 37 W. Table 6-3 shows the typical technology parameters, including the metal layer pitches.

Figure 6-10 shows the die photo. The measured power dissipation at 300 MHz is 37 W. There are 1600 area C4 and 448 off-chip signal I/Os. Dedicated thin-oxide capacitors of 102 nF are provided for on-chip decoupling [69]. Combined with the "built-in," non-switching well-to-substrate and diffusion-to-well capacitances, the total on-chip decoupling capacitance is about 200 nF [69].

The power distribution supports an average DC voltage drop of 23 mV. The Delta-I current transients were managed by including additional on-chip decoupling capacitors around large noise sources, such as the off-chip drivers, clock buffers, and on-chip drivers with large loads. Since a large amount of switching capacitance occurs in the dataflow stacks, decoupling capacitors were also placed under the wiring tracks.

The thin-oxide capacitor features a "built-in" fuse mechanism whereby weak spots between M1 and contact are used to blow connections to V_{dd} and ground in the presence of a large current resulting from oxide defects. Each capacitor has a gated NFET control device with an external *decap_enable* pin for leakage current measurement during testing.

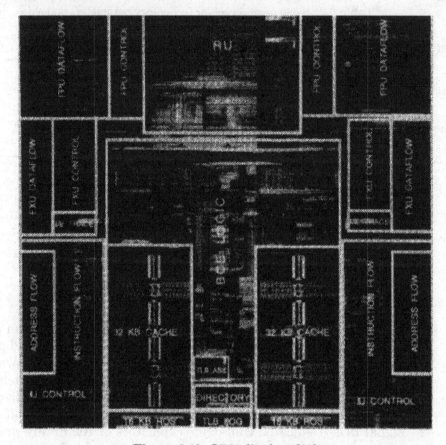

Figure 6-10. S/390 die photo [69].

Figure 6-11 shows the decoupling capacitor cell that fits under the data flow wiring tracks. The cell is double bit-pitch wide (43.2 μm) and 14 tracks tall (25.2 μm). Two out of the 14 horizontal wiring tracks are specially blocked for the decoupling capacitor wiring so the capacitor can fit right under the wiring tracks. A low-resistance layout of the capacitor cell provides a fast time constant of about 85 ps.

6.5   SUN SPARC 64B MICROPROCESSOR

This die with 750 I/O signals and 1735 power bumps is flip-chip-attached to a multilayered ceramic land grid array package [70]. Figure 6-12 shows the die micrograph of the chip [70]. The pack-

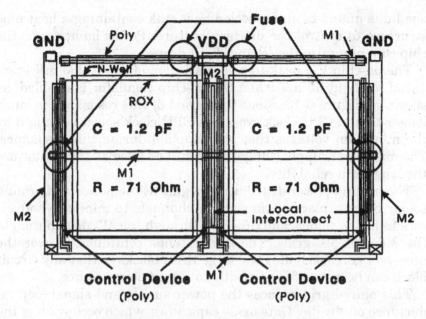

Figure 6-11. Decoupling capacitor [69].

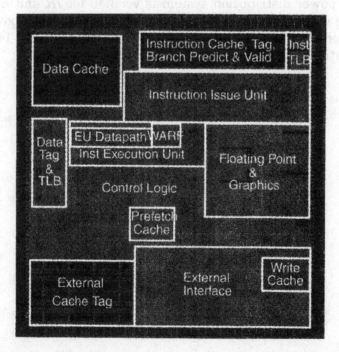

Figure 6-12. Die micrograph of Sun SPARC 64-bit microprocessor [70].

age lid is mated to an air-cooled heat sink containing a heat pipe structure to control the die temperature. Power bumps over the chip core minimize the IR and di/dt drops.

The on-chip V_{dd} peak-to-peak variation of about 260 mV is reduced to about 60 mV when the on-chip regulator is enabled, as shown in Figure 6-13. Since the period of this resonance is much longer than a CPU clock cycle, the CPU clock speed is limited by the minimum voltages that are supplied during this resonance. The maximum supply voltage must still be fixed at 1.6 V to assure the long-term reliability.

The on-chip power distribution begins at the power and ground solder bumps, placed primarily in channels to minimize soft errors from the solder, and proceeds through the M7 distribution to the M6 and M5 grids. The grid extends continuously over the processor core, excluding the large RAM blocks so that any circuit block can be connected vertically to a good power source.

This paired grid reduces the power supply and signal loop inductance on the die. Gate oxide capacitors, which occupy all of the unused silicon area under the wiring, connect to the power grid to increase the on-chip bypass capacitance by 220 nF.

The power distribution system is verified for IR and EM compliance using a Cadence tool [60]. This tool checks the power distribution on both static and dynamic modes. Figure 6-14 shows one static simulation result for the IR drop plot.

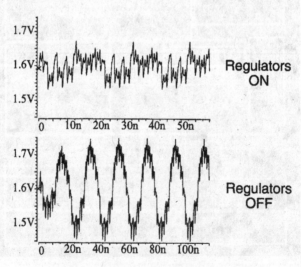

Figure 6-13. Supply voltage noise [70].

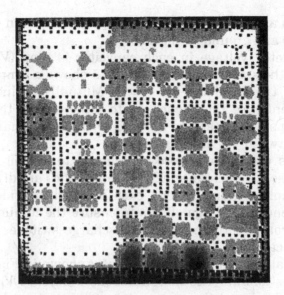

Figure 6-14. Full-chip *IR* drop plot [70].

This simulation was done after the core was attached to the pad ring and the result shows a black region in the bottom right of the die. This large *IR* drop being highlighted is where the power supply connections between the core and the pad rings are incomplete. A hook-up is added here later to fix this *IR* drop problem.

Voltage regulation requirements of each generation of microprocessors are more critical as the on-chip voltage decreases and the AC current increases. Distributed thin-oxide capacitors are used for supporting instantaneous current variations within the die, but are insufficient to compensate for the tank circuit formed by the parasitic LC in line with the supply distribution.

Simulation shows nearly an order of magnitude increase in supply network AC impedance seen by an internal gate at resonance. This resonant frequency is much lower than the system clock frequency but can limit the speed performance. A special voltage regulator circuit is placed 99 times to reduce the resonance from the board to the package to the chip.

The voltage regulator circuit increases the charge stored or delivered by a given amount of added decoupling capacitors by actively increasing the voltage across the capacitor's terminals. The operation is done by stacking fully charged equal value capacitors in series as a voltage multiplier to supply charges in the on-chip power (V_{dd}) and ground (V_{ss}) grid.

The depleted voltage in each capacitor is then ($V_{dd} - V_{ss}$)/n, where n is the stack height. Figure 6-15 shows a simplified block diagram of the regulator for $n = 2$. Mutually exclusive CMOS switches configure the capacitors to either be in the charging phase when shunting across $V_{dd} - V_{ss}$, or in the discharging phase in series between V_{dd} and V_{ss}.

The sizes of the capacitors are chosen to exhibit the proper equivalent series resistance ESR. The switches are driven by two sets of complementary drivers, each of which provides two outputs with enough voltage offsets to ensure the minimal crowbar leakage through both charge and discharge switches during the switching activity.

The operation of the voltage regulator shown in Figure 6-15 is described as follows. The instantaneous difference V_{inst} between V_{dd} and V_{ss} begins at the same value as the average $V_{dd} - V_{ss}$. In this condition, N2 and P2, the shunt switches, are weakly on with gate-to-source voltages of ($V_{dd} - V_{ss}$)/2 each, whereas N1 and P1,

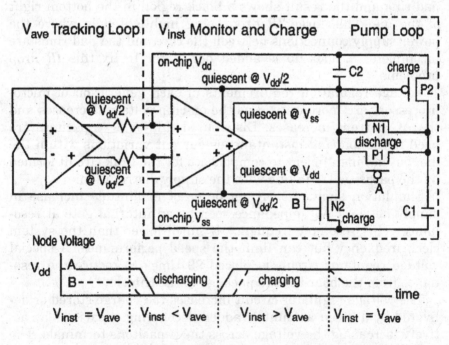

Figure 6-15. Block diagram of voltage regulator [70].

the series switches, are completely off. Then V_{inst} drops, causing node B to fall, cutting off N2. Slightly later, node A falls, turning on P1. This changes C2 from being in shunt with C1 to being in series. Similarly, the mirror devices, P2 and N1, are being cut off and turned on, respectively. This allows the series-connected C1 and C2 to discharge into the power grid, which forces V_{inst} up. In the next time section, where $V_{inst} > V_{ave}$, node A rises, cutting off P1, and then node B rises, turning on N2. Similarly, N1 turns off and then P2 turns on. This switches C1 and C2 into the shunt mode, allowing them to be charged by V_{inst} and forces V_{inst} to drop. Once $V_{inst} = V_{ave}$, node B returns to $V_{dd}/2$, which returns the circuit to the weakly charging mode.

The switched capacitors are enhancement mode MOSFET devices, laid out in a waffle-type structure to maximize capacity [70]. The regulators are evenly distributed across the chip in 99 instances, which are directly hooked up to the main global power grid.

Care has been taken in shielding sensitive signals and in managing high-current-density paths. The regulators are placed underneath the global routing channels to reduce the layout area impacts.

6.6 INTEL IA-64 MICROPROCESSOR

This microprocessor implements a highly parallel execution core, while maintaining binary compatibility with the IA-32 instruction set [71]. The processor contains 25.4 million transistors, and is fabricated in a 0.18 μm CMOS process with six metal layers using C4 or flip-chip assembly technology in an organic land grid array.

Table 6-4 shows the process technology used in the manufacturing of the processor. Figure 6-16 shows the die micrograph for this processor and Figure 6-17 shows the architecture [71]. Four 1MB L3 cache chips are connected to the processor die by a core-speed backside bus (BSB).

Table 6-4. 0.18 μm process technology [71]

Poly	M1	M2	M3	M4	M5	M6
0.48 μm	0.60 μm	0.72 μm	0.72 μm	1.45 μm	1.80 μm	2.00 μm

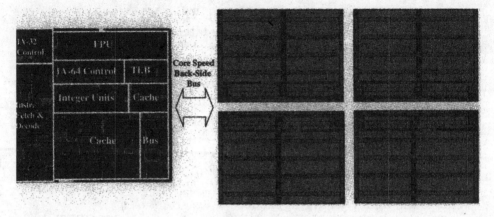

Figure 6-16. Die photograph of Intel IA-64 microprocessor [71].

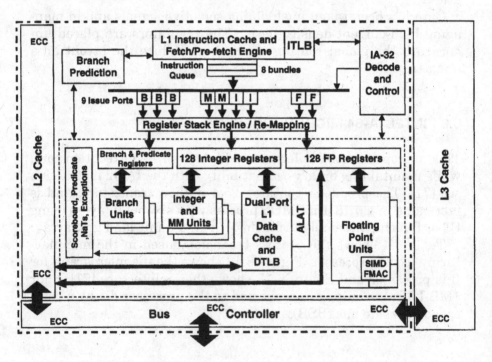

Figure 6-17. Architecture of Intel IA-64 microprocessor [71].

All these components are packaged in a cartridge optimized for double-sided motherboard mounting, as shown in Figure 6-18. The processor has fifteen execution units, including four integer and two floating units. The processor includes three levels of cache organized in the hierarchical manner. The L1 and L2 caches are integrated on the die. The L3 cache contains up to 4 MB of custom-designed on-cartridge memory and is connected to the processor die by a dedicated 128 bit BSB source synchronous interface.

Power is delivered from the voltage converter to the processor cartridge through a separate connector that provides significantly lower impedance compared to traditional power delivery using pins through the motherboard socket. The chip-level power distribution consists of a uniform M6–M5 grid with C4 power and ground bump arrays.

This grid has the power and ground lines finely interspersed with signal traces to reduce the inductive crosstalk, i.e., a very wide power or ground line is composed of multiple thin lines of the V_{dd} and V_{ss} in order to reduce the inductance talk or the switching current returning paths.

The on-die decoupling capacitors are placed in the proximity of the high di/dt switching circuits, as well as in all the routing channels. The total on-die decoupling capacitance is about 800 nF in this microprocessor [71]. In addition, on-package decoupling capacitances have been added to reduce the synchronous switching noise from the I/O buffers.

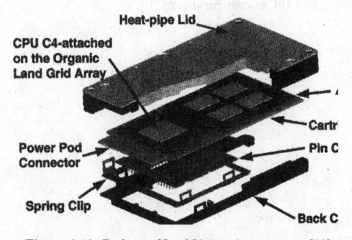

Figure 6-18. Package of Intel IA-64 microprocessor [71].

This microprocessor allows the use of clock gating to reduce the average power without any loss in performance. Figure 6-18 shows the internal structure of the processor cartridge. The processor is C4, attached to a multilayer organic land grid array (OLGA) package, which is soldered to the base cartridge substrate. Inductive signal return current loops are minimized by proper placement of return vias for image currents propagating in the reference planes inside the multilayer package.

6.7 SUMMARY

With microprocessor frequency continuing to rise and supply voltage continuing to decrease, the power delivery system remains very challenging in microprocessor design. The C4 or flip-chip package, with area solder bumps, is used in modern microprocessor chips. The dense power grid in multiple metal layers is used to achieve low-resistance delivery of power inside the die.

To prevent di/dt noise collapse for the circuit functions, a large amount of decoupling capacitors have been used in the chips. Package design with decoupling capacitors is essential to provide the lower voltage drop; multiple power and ground planes are used for this purpose.

In addition, the voltage regulators of microprocessor chips have been moved into the die to stabilize the increasingly reduced on-die power supply voltage, as in the design example shown in the Sun SPARC 64-bit microprocessor [70].

7

PACKAGE AND I/O DESIGN FOR POWER DELIVERY

The power delivery performance of a VLSI system depends not only on the on-chip power network, but also on the system-level power distribution, including the package options and board power planes. The voltage drop and power noise are influenced by the chip, the package, and the entire board. Each of the components in the system will contribute to the voltage noise as a whole. Therefore, the package options and I/O design for power supplies are important in the VLSI power network design.

This chapter is organized into six sections. Section 7.1 describes the flip-chip package technology. Section 7.2 discusses the simultaneous switching noise for off-chip drivers. Section 7.3 provides a case study of how to evaluate the package technology and metal options in a high-performance microprocessor [76]. Section 7.4 discusses microprocessor power noise measurement techniques. Section 7.5 describes the I/O pads for power and ground supplies to the chip. Section 7.6 summarizes the chapter and also highlights some thoughts on the chip and package codesign concept [81, 82].

7.1 FLIP-CHIP PACKAGE

The length of the electrical connections between the chip and the substrate can be further reduced using flip-chip or C4 technology. This technology is achieved by distributing the I/O solder bumps over the die, flipping the chips over, aligning them with the contact pads on the substrate, and connecting the solder bumps between the chip and package to make connections.

This saves silicon area and increases the maximum number of I/O and power/ground terminals available with a given die size. This package also provides more efficiently routed signal and power/ground interconnections on the chips. Therefore, modern high-speed chips and microprocessors use this flip-chip technology to achieve high speed and lower power noise.

For example, the 450 MHz RISC microprocessor from Motorola has a chip footprint with a total of 794 C4 or flip-chip pads [72]. Two hundred and sixty-six pads are used for 64-bit bus transfer, 64-bit L2 interface, and control. The remaining C4 pads are used for power and ground and possible extension to 128 bit bus transfer and L2 interface options.

The 1.8 V V_{dd} and ground C4 pads are distributed over the core of the chip to reduce the voltage drop and feed the internal power structure. The signal I/Os are distributed around the periphery to reduce the wiring congestion in the package substrate and to isolate the ESD structures from the internal circuits.

L2 cache interface C4s are placed along the left side (bits 0–63) and bottom (bits 64–128) of the chip. This allows for an optimal multichip module design of this processor, with two SRAMs using the 360-pin solution. The data transfer signals are on the right side of the chip, and the address/control signals are at the top.

A total of 236 V_{dd} and ground C4 pads are used for the internal 1.8 V core and supply 1.8 V power to off-chip I/O drivers and receivers, 55 V_{dd} and ground C4s for the external L2 interface, and 73 V_{dd} and ground C4s for the external bus transfer address and control [72].

Flip-chip connection technology as the first level chip-to-package connection option traditionally is regarded as being the controlled collapse chip connection (C4) process, which was originated by IBM [73].

Figure 7-1 shows the schematic, which is a bare IC device flipped upside down with its active area or I/O side attached to a substrate via a connecting medium. The device may be any of the substrates providing an interconnection network between the flipped active device and other active, or even passive devices, such as the decoupling capacitors.

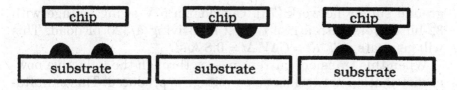

Figure 7-1. Flip-chip package [73].

There is another feature unique to having the active side of the chip face the top of the interconnecting substrate. Since the I/O pads on the chip also are fabricated on the active side, the layout of these pads easily can be expanded into an array covering the entire inner area of the chip, rather than being confined to the perimeter.

Area arrays I/Os in the flip-chip package offer a way of increasing I/O density. For a chip size of 5 mm and a constant I/O pad spacing of 100 μm, a perimeter array could accommodate about 200 I/Os, whereas an area array could accommodate about 2000 I/Os, a tenfold increase.

Only the flip-chip configuration provides the ability to achieve higher I/O density without decreasing I/O pitch. Flip-chip bonding also offers the shortest possible leads with the lowest inductance, maximizing the operating frequency.

Table 7-1 shows the typical values of the lead inductance and capacitance in various chip package choices. The solder bump provided by flip-chip technology has the lowest inductance and lowest capacitance, compared to wire bonding and TAB technologies [74].

7.2 SIMULTANEOUS SWITCHING NOISE (SSN)

When a number of off-chip loads are switched simultaneously in a digital system, a current change is produced in the power and

Table 7-1. Typical values of lead inductance and capacitance [74]

Package Technology	Capacitance (pF)	Inductance (nH)
Wire bonding	0.5	1–2
TAB	0.6	1–6
Solder bump	0.1	0.01

ground supply network [73]. Consider a 5 V swing voltage with 32-bit drivers with a rise time 2 ns driving a 320 pF load. This will generate a $di/dt = C\Delta V/\Delta t = 0.8$ A/s.

When this transient current passes through the inductive power distribution network, a noise voltage is produced. This simultaneous switching noise is sometimes referred to as *ground bounce*. The switching noise can result in a number of problems if not handled correctly.

The noise appears at the output of what were intended to be quiet off-chip drivers. This noise appears at the inputs of the connected receivers. The changes in the internal chip supply voltage make the circuits operate more slowly, and thus increase the delay in switching drivers.

Overshoots and undershoots might also appear in these drivers. For on-chip circuits acting as input gates, the simultaneous switching noise acts to reduce the effective noise margin at the inputs. For on-chip memory devices, such as latches, large amounts of the ground rail and power rail noise might cause false changes in the logic state. In the first order, the noise generated by the simultaneous switching of N output drivers can be calculated as follows [74]:

$$\Delta V = N \cdot L_{\text{eff}} \cdot di/dt \qquad (7\text{-}1)$$

where L_{eff} is the effective inductance of the power and ground connections, di/dt is the peak rate of the change of the currents for each driver, and N is the number of drivers used during the switching. di is the current demand of each driver during the switching event, and dt is the rise and fall time of the signal.

In reality, the ΔV does not increase linearly with L_{eff} or N, because any increase in ΔV will slow down the circuits and reduce the di/dt. The effective inductance L_{eff} is primarily a function of the package design. Reducing L_{eff} requires minimizing the inductances of the power and ground distribution networks and also the use of the decoupling capacitors.

The decoupling capacitor placed between the power and ground pins of each chip can act as a local source of the charges during the switching events, so that not all of the switching current has to be supplied from the system ground to minimize the local change in voltage. Figure 7-2 shows the equivalent circuit model for a CMOS output driving a capacitance [74].

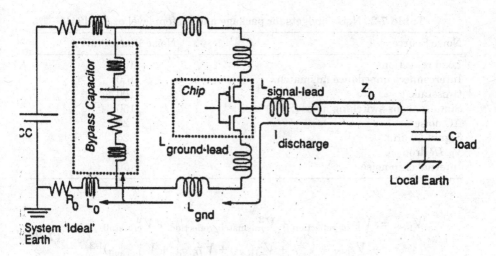

Figure 7-2. Electrical modeling of a package power distribution network [74].

A couple of inductances are included in this model: the inductance of the ground lead in the chip attachment and the inductance of the ground plane or wiring between the chip attachment and the decoupling capacitors. The parasitic inductance and capacitance associated with the decoupling capacitor are also shown in this figure.

To minimize the L_{gnd} and L_0, the ground and power planes are used in the package design. The decoupling capacitors on the package should be placed close to the chips, since at high frequencies it is important to minimize the parasitic R and L of the decoupling capacitors. More leads or I/Os assigned to the chip are preferred to reduce the inductance.

There are several other sources of noise that must be considered in the package design [74]. Solutions include the use of multiple power supply planes or using a ceramic substrate base with thick-film ground and power planes within it. Table 7-2 shows the relative noise budgets for each noise source, including reflection noise, crosstalk noise, and simultaneous switching noise [74]. These noise budgets include two different types of reflection noise: reflection from loads and reflection due to mismatches between different transmission lines.

The simultaneous switching noise refers to the noise at the outputs of the quiet drivers when they are grounded. The root sum of squares of the different noise voltages are calculated as follows [74]:

Table 7-2. Noise budgets for package and system level at 6°C [74]

Noise Source	Noise Budget (mV)
Load reflections	100
Interconnect impedance mismatch	100
Crosstalk	100
Simultaneous switching noise (SSN)	150
AC noise	25
Signal IR drop	25
V_{cc} IR drop	14
Internal chip noise	50

$$V_{\text{RSS}} = (V^2_{\text{load\_reflection}} + V^2_{\text{mismatch\_reflection}} + V^2_{\text{crosstalk}}$$
$$+ V^2_{\text{SSN}} + V^2_{\text{AC}} + V^2_{\text{IR-sig}} + V^2_{IR\text{-}Vcc} + V^2_{\text{thermal}})^{1/2} \tag{7-2}$$

The parameters in the package model, such as the simultaneous switching noise model, as shown in Figure 7-2, are provided by the package vendors. In addition, the transistor models of the I/O drivers are also included in the simulation model. An electrical study will provide the amount of decoupling capacitance and package layers design guidelines.

In addition, after the layout of the package layers has been done, the extraction of the RLC parasitic is provided, based on CAD tools, and then the circuit simulation is done to measure the performance of the package design, especially for the simultaneous switching noise against the required budgets.

The simulation conditions are set up correctly to model the circuit operation environment. Any deviations from the simulation will be reported as a possible drawback in the design and improvements will be adopted; for example, adding more decoupling capacitors or using additional power and ground planes on the package.

Figure 7-3 shows a simulation model, with the simulation conditions switched within 2.5 ns, and with a ramp-up and ramp-down peaking at 1.25 ns. Table 7-3 provides the assumptions made on the package and chip parameters. The ground or V_{ss} inductance and resistance parameters reflect the V_{dd} path parameters. Four different package types were investigated with the parameters, as shown in Table 7-3.

The simulation model of the package-level power network can be simplified to a tank circuit, as shown in Figure 7-4. L and R are the lumped parasitic inductance and parasitic resistance of the

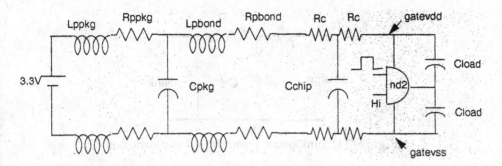

Figure 7-3. Simulation model of package performance [75].

power distribution network from the voltage regulator or the voltage source to the chip. C_d is the total capacitance at the inputs of the chip, including the added decoupling capacitors on-package and on-chip. The resonance frequency of the tank circuit is given as follows:

$$f = \frac{1}{2\pi\sqrt{LC_d}} \tag{7-3}$$

The resonance quality factor Q determines the impedance of the network at the resonance frequency as follows:

$$Q = \frac{\sqrt{LC_d}}{R} \tag{7-4}$$

For the design improvement, we can increase C_d so that the resonance frequency f is very small compared to the operational frequency range, and Q is small. We can also decrease the package inductance L to the extent that f is very large. We can achieve

Table 7-3. Simulation model parameters of four packages [75].

	Lp/g Package	Rp/g Package	Lp/g Bond	Rp/g Bond	Rc
Package A	180 pH	1.0 mΩ	180 pH	1.0 mΩ	2.5 mΩ
Package B	80 pH	1.0 mΩ	90 pH	1.0 mΩ	2.5 mΩ
Package C	67 pH	1.0 mΩ	74 pH	1.0 mΩ	2.5 mΩ
Package D	55 pH	1.0 mΩ	30 pH	0.5 mΩ	2.5 mΩ

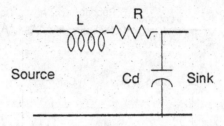

Figure 7-4. Tank circuit model for power distribution.

very high frequency with the flip-chip package and use both the package-level and on-chip decoupling capacitors.

The degree of ground bounce depends on multiple factors, such as the total current inputs to the chip, the clock delay and skew, and the switching activity factor. With reduction in clock delay and clock skew, the higher harmonic components will become stronger, and the most acceptable design technique would be staged decoupling, with both on-package and on-chip decoupling to reduce the package resonance to a small value.

With no additional on-package decoupling capacitance, it requires a very large on-chip capacitance to decouple the package inductance. As the power and ground inductance decrease to meet the simultaneous switching noise reduction requirement, the required on-chip capacitance becomes larger. On-package decoupling capacitors should be used to decouple the package inductance.

The value should be high enough to make the first resonant frequency and the resonant impedance sufficiently low. The resonant frequency, as specified in Equation (7-3), should be four to five times smaller than the clock frequency.

The following parameters should be considered when predicting the high-frequency ground bounce:

1. The chip current demand and clock skew
2. Package *RLC* parasitic
3. Chip RC parasitic of the power and ground network
4. The number of gates, the activity factor, and the average loading on each gate to estimate the on-chip capacitance

In the worst case, when there is no on-package capacitance or large power and ground inductance, the high-frequency bounce on either power rail is roughly determined, based on the above factors [75].

The ground bounce predominately observed in the time domain analysis is referred to as low-frequency bounce; it occurs with a frequency of f_{clk} and $2f_{clk}$ [75]. The magnitude of the low-frequency bounce may be conservatively estimated by the following equation [75]:

$$\text{Bounce}(f_{clk}) = \frac{P_{core}}{3.3 \cdot Zin(f_{clk})} \qquad (7\text{-}5)$$

Here P_{core} is the power dissipation due to the core gate, including the flip-flops. As the power dissipation increases, it becomes necessary to decouple at very high frequency.

With less on-chip decoupling, it is important to reduce the chip-to-package inductance with integrated decoupling, along with large high-performance on-package decoupling. The bounce magnitudes observed in the simulations are less than 70% of the values predicted by the above equations [75].

The delay derating factor for the ASIC standard cell library is $Kv = 1.03$ for a 160 mV reduction in the supply voltage, or 3% increase in the delay for a 5% reduction in the voltage [75]. For 320 mV peak bounce on both V_{dd} and V_{ss}, the delay penalty is 6%, approximately the dynamic effect of the bounce with an average effect.

For a critical path in a 100 MHz system, if only a 5 ns delay with gate and loads is produced by this bounce, the delay penalty is 300 ps [75]. Unless the low-frequency bounce is designed within a controlled limit, the effect on chip power consumption may be noticeable. This results are from the fact that the power consumption from the ground bounce affects all gates in the chip.

The simulations indicate a significant variation in the power dissipation. For example, for a very high performance package with no on-package decoupling, the power dissipation may vary from 13.0 W to 18.9 W [75]. One effect of the power consumption from low-frequency bounce is that it is dependent on the relative position of the clock frequency and resonance frequency.

A study methodology for the ground bounce and decoupling capacitance has been proposed as follows [75]:

1. Obtain the current requirement of the chip based on real analysis or based on some known parameters, such as gate

count, activity factor, clock delay, clock skew, current peak, current width, etc.

2. Obtain the package RLC characteristics and the chip RLC characteristics.

3. Design the preliminary decoupling network based on the equations.

4. Obtain the frequency domain characteristics of the decoupling network through SPICE simulation and modify the network using the measurement information.

5. Obtain the dominant frequency components of the current waveform and extract the magnitude of current waveforms in the desired frequencies. The desired frequencies are commonly f_{clk} and $2f_{clk}$.

6. Check the decoupling condition at high frequencies ($5f_{clk}$ to $8f_{clk}$) to eliminate the high-frequency bounces.

7. Check the decoupling condition at low frequencies all the way to DC. This is particularly important if specialized megacells like memories are used, which can excite very high harmonics.

8. Set a targeted bounce number and modify the decoupling network.

9. Distribute the on-chip capacitance according to the localized current demand within the chip.

10. Distribute a power and ground network on the chip to minimize the localized bounce.

11. Verify the on-chip power distribution for local hotspots after the layout. This will require modeling and extraction capabilities for on-chip power distribution parameters like resistance, inductance, and capacitance.

To address the high-performance and high-integration applications, flip-chip technology with integral power and ground technology should be used, along with the on-chip decoupling provided by the gate capacitance.

Lower-cost packaging solutions may be available for low-integration–high performance and high-integration–low-performance applications, but the design should go through the power distribution methodology, integrating both the on-chip and on-package decoupling. This also emphasizes the need for a chip design flow

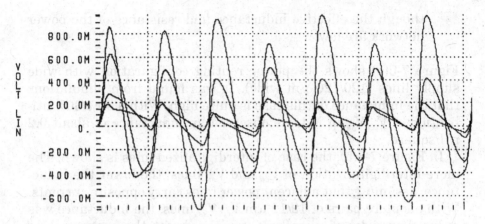

Figure 7-5. V_{ss} ground bounce in a flip-chip package [75].

that includes the package. Figure 7-5 shows the simulation result of the V_{ss} bounce noise for a flip-chip package [75].

7.3   CASE STUDY OF A MICROPROCESSOR-LIKE CHIP

The purpose of this case study is to analyze the power network on a microprocessor-like die for several technology options. The study is based on a distributed model of the chip, with current sources representing the active circuitry. The model is tested for normal, power saving, and power peak modes. The die size is about 17×17mm^2 and the power supply is about 2.5 V, with an average current of 12.5 A for the average power of 31 W [76].

The power network was known to be a significant problem in terms of both metal utilization and voltage drop in the center of the die. There are several options considered in this study as follows [76]:

1. Thick M4 with wire bond.
2. Routing most of the V_{ss} through the substrate to reduce crowding on M4 and improve routability as well as average voltage in the center.
3. Wire bond with M4/M5.
4. Using C4 with M4. In C4 technology, the power is routed through the package. The M4 utilization is very low, al-

though the effective inductance and resistance of the power network are low.

Figure 7-6(a) shows the power routing configuration with wide supply lines (120/120 μm wide). It was found from simulations that the inductance of this case is quite high. With a 30%/30% utilization of V_{cc}/V_{ss} in M4, we observed an inductance of about 0.2 nH/square.

In Figure 7-6(b), the case of interdigitalized lines is shown. The more interdigitalization in V_{cc} and V_{ss} lines, the lower the inductance, assuming that adjacent power lines carry opposite currents. With about 10 pairs of 12/12 μm V_{cc}/V_{ss} lines, the inductance was reduced an order of magnitude, compared with the power routing line widths shown in Figure 7-6(a), to the 0.02 nH/square.

The package for a wire-bond case is shown in Figure 7-7(a). Assuming that discrete low-inductance capacitors used in the package and a total of 300 bond wires for V_{cc} and V_{ss}, the total package inductance is 114 pH per side, and the bond-wire inductance causes 65% of the total inductance.

The power in the package is supplied only from two sides. From the process point of view, it is easier to make the last metal thick-

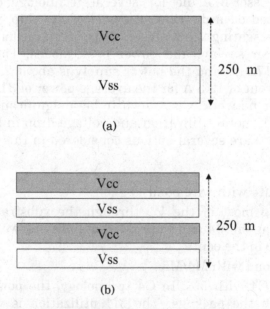

(a)

(b)

Figure 7-6. V_{cc} and V_{ss} line configurations [76].

Top View:

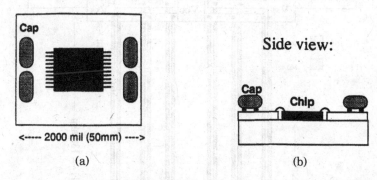

<----- 2000 mil (50mm) ---->

(a)

Side view:

(b)

Figure 7-7. Wire-bonding package with decoupling capacitors [76].

er than the others; and from a routing point of view, it is preferable to have most of the power routed in the thickest, uppermost metal layer.

The bonding wire is 2.7 m long with radius 12.5 μm and pitch 125 μm. Each bonding wire has an inductance of approximately 72 pH. The discrete capacitor lead has an inductance of 15 pH. The V_{ss} and V_{dd} planes have about 27 pH inductance, and the total wire-bonding package inductance is about 114 pH (72 + 15 + 27).

The C4 package is shown in Figure 7-8. We now have 50 μm long solder balls with radius 64 μm and pitch 250 μm. The total number of C4 balls is about 3000. The C4 inductance has been reduced to be negligible. The C4 has completely removed the package-to-chip bottleneck. We also benefit from the four-sided supply of the C4 due to the power planes in the package below the chip.

The effective package inductance has been reduced from 57 pH in the wire-bond case to 10 pH in the C4 case. This can be further improved by placing the package decoupling capacitors closer to the chip, and by using a large number of on-chip decoupling capacitors.

The power routing is assumed to be in Manhattan structures in M3 and M4. The initial estimates are based on average currents in a uniformly distributed load on the chip. These values will then be tested and refined by using the distributed power supply model.

In the following, we will estimate the IR drop in the wire-bonding technology in which the pads are located on the boundary of

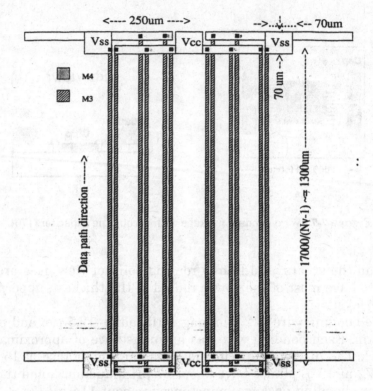

Figure 7-8. C4 power routing configuration [76].

the chip to feed the middle of the chip. Both M4 and M5 are assumed to be 1.8 μm thick and about 21 mΩ/square. This is an optimistic assumption, since only the last layer can be made significantly thicker than the other layers. M3 is assumed to be 0.8 μm thick and about 47 mΩ/square.

The average current drawn by the chip is about 12.5 A and the supply voltage is about 2.5 V. We assume in all cases 30% for V_{cc} and 30% for V_{ss}. M3 is used for equalization of about 5%/5% V_{cc}/V_{ss}. The effective resistance of M4 for V_{ss} or V_{cc} is increased by 70 mΩ/square [76].

The average voltage drop can now be calculated by considering uniform current injection from one side. The current is reduced by two times, the resistance is only 0.5 square, and the current is reduced linearly from the edge of the chip to the middle as follows:

$$V_{\text{drop}} = (I_{\text{tot}}/2) \cdot (R_s/2) \cdot (1/2) \qquad (7\text{-}6)$$

where the I_{tot} is the total current consumed by the chip and R_s is the metal sheet resistance. For the case of the interdigitated power supply in M4 with the V_{cc}/V_{ss} metal widths of 30 μm/30 μm, as shown in Figure 7-6(b), the power is supplied only from two sides and only M4 is used to carry the average current.

Based on Equation (7-6), for this case, the voltage drop is calculated as: $V_{cc\_drop} = 6.25 \cdot (52.5e - 3/2) \cdot 0.5 = 82$ mV. The average drop in the V_{ss} is dependent on the number of substrate taps. The number of taps is determined by peak noise considerations, so the average voltage drop will be small.

On average, we could get the $V_{cc} \sim V_{ss} = 130$ mV, so we could achieve a good routability with only 40% V_{cc} in M4 and a tolerable average voltage drop. But this power routing configuration with the wire-bonding package has high inductance and, therefore, it has a high switching noise drop across the package and chip.

We consider the second option of the metal routing for V_{cc} and V_{ss} using the wire-bonding package. The M4 V_{cc} and V_{ss} are 30/30 μm in width and the M5 V_{cc} and V_{ss} are also 30/30 μm in width. We assume that the M4 and M5 have the same metal thickness. We also assume that the power supply is from all four sides of the chip, so the inductance will be reduced.

We can roughly estimate that the voltage drop from the chip side to the middle is reduced to 220 mV/2 = 110 mV, and the routability is also improved significantly with the fifth metal layer (M5) added for the power routing [76].

The C4 power distribution is quite different. The resistance in the package plane is only 2.36 mΩ/square, so the voltage drop in the package from the edge of the die to the center, assuming a uniformly distributed current injection to the chip, is about (12.5 A/4) · Rs/4 ≅ 2 mV [76].

One suggestion is to place the power routing on the package instead of on the chip. The maximum number of solder bumps on the $17 \times 17 \text{mm}^2$ chip with a minimum pitch of 250 μm is $17^2/0.25^2 \cong 4600$.

Since the landing pad of the solder ball is 70×70 μm, the total area used, if we use a maximum number of solder balls, is $4600 \cdot 0.07^2 = 23$ μm^2, which is about 8% ($23/17 \cdot 17$) of the chip area. By using about half of the solder bumps for power and ground, we need little local routing in M4/M3 from the solder bumps. In addition to reducing the inductance, the C4 technology also reduces the on-chip power routing significantly.

The following is one option we will discuss to use C4 with M4 and M3 for local power distribution. Since the area between the bumps in the horizontal direction is not available for signal routing, we might as well use the minimum pitch solder balls in M4 with alternate V_{cc} and V_{ss} in order to minimize the inductance.

If we assume 15 rows of M4 solder bumps for the whole chip, and 30 μm/30 μm for V_{cc}/V_{ss} in each row, the resistance is 2/mΩ · 17,000/15 · 30 = 800 mΩ/square. By reducing the horizontal distance by N, we reduce the current injected in each section by N and the resistance to the center of each section by N.

We can neglect the voltage drop in the package, so the voltage drop will thereby be reduced by N^2. With a solder bump at each 250 μm, we have a V_{cc} each 500 μm => N = 17,000/500 = 34. Figure 7-8 shows the C4 package power routing configuration.

A full-chip and package model of the power distribution network is built, as shown in Figure 7-9 [76]. This model is used to simulate the effect of different metal utilizations and packages more accurately. The 25 or 5 × 5 elements in the center model the chip core. Separate current waveforms can be injected into each of these elements in order to model real chip blocks with different activities.

Around the core, there are five package elements at each side to model the wire bond or C4 package. Part of the C4 package model is also included in each core element, since the solder ball bumps can be placed anywhere on the die. The pins of the package are assumed to have ideal V_{cc}/V_{ss} potentials.

The core element consists of three main elements as follows:

1. The current source for V_{cc} and V_{ss}. The current waveform can be injected between the local V_{cc} and V_{ss} power supplies.
2. The decoupling capacitances, with the modeling of parasitic capacitance and an explicit decoupling capacitance.
3. Power network metal RL modeling.

The RL branches in the simulation model show the on-chip V_{cc}, V_{ss}, and substrate per unit resistance and inductance, as well as the C4 V_{cc} and V_{ss} package planes. Figure 7-10 shows the power I/O package model. It uses separate inductors and resistors for C4 and bond wire package models [76]. It includes both C4 and wire-bonding models in one simulation model. In order to switch

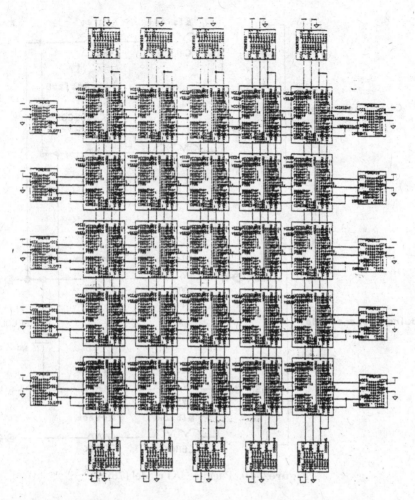

Figure 7-9. Full-chip and package model of power distribution network [76].

from C4 to wire bonding, we can change all the C4 resistance values to 10 kΩ so that there will be negligible current in the C4 network.

Both experimental results and RC extraction of all the different parasitic components of the on-chip capacitance suggest that the total effective on-chip decoupling capacitance on the previous microprocessor using the old process is about 40 nF. In the new process, which has the 0.8 scaling factor from the previous microprocessor, the main assumptions for the capacitance in the new microprocessor in this experiment are as follows [76]:

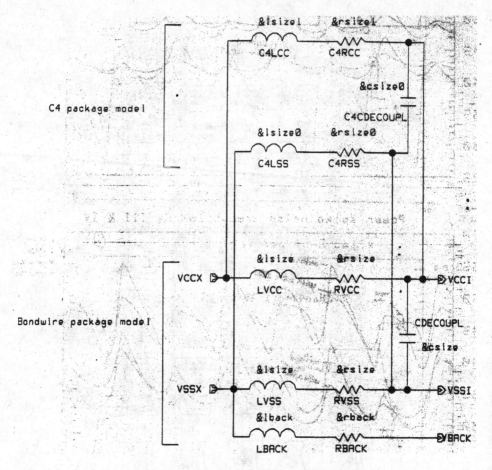

Figure 7-10. Package I/O model [76].

1. The n-well and diffusion capacitance increase due to smaller reverse bias by 1.3 times, which is due to higher doping of ~1.25 times in the new process.
2. The gate oxide increases by 2.2 times due to the gate thickness.
3. The metal capacitance increases by 1.4 times due to extra-level metal and smaller pitch.

The total parasitic capacitance is estimated to be about 75 nF with an uncertainty of about ±15 nF. The RC components of the decoupling capacitances are also included in the simulation model. It turns out that the n-well decoupling capacitance is not very effective in absorbing short spikes, due to the high lateral resistance in the well.

Figure 7-11 shows the average noise with and without the 100 nF on-chip decoupling capacitors. Obviously, the noise performance is better with the additional decoupling capacitors. However, the noise without the additional decoupling capacitors is not that severe, so the intrinsic parasitic capacitance does help significantly.

In order to find the exact requirement of the on-chip decoupling capacitance, a better knowledge of the clocking strategy, the power saving requirements, and bus protocol is necessary. In the power network noise simulation, the switching currents are modeled on the power grid model. The modeling of the switching currents is the key to the power noise results.

Figure 7-12 shows the current waveform used in Figure 7-11's result. It uses a waveform peaking at the beginning of the cycle with the rising edge of the inserted clock and falling off to 50% and 20% of the peak at the middle and the end of the cycle, respectively.

The average current is 500 mA and the peak is about 800 mA, and with 25 cells in the full-chip modeling this results in current

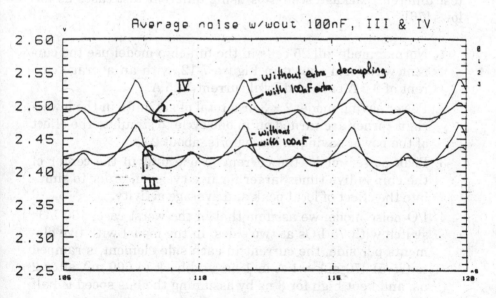

Figure 7-11. Simulation waveforms for V_{cc} noise [76].

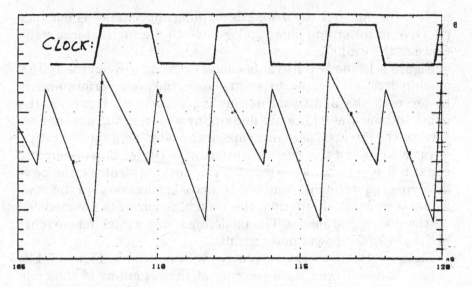

Figure 7-12. Switching current waveforms injected into each cell [76].

consumption on the chip of about 25×500 mA = 12.5 A [76]. The I/O cells may use different current waveforms from the core cells.

By using the current waveforms shown in Figure 7-12, we can test different package scenarios using different test cases as follows [76]:

1. Normal mode: all 25 cells in the full-chip model use the current waveform shown in Figure 7-12, with an average current of 0.5 A each and total current 12.5 A.

2. Power saving mode: 2×3 or a total of six units in the lower-right corner are turned off in one cycle to simulate the effect of the power saving in large units, about 24%.

3. Peak power mode: the current in one unit in the center of the chip is five times larger for five cycles in order to simulate the effect of local peak and average activity.

4. I/O noise mode: we assume that in the worst case, 150 I/Os switch with 75 I/Os at two sides. In the model with five elements per side, the current in each side element is ramped to $(75/5) \cdot 70$ mA = 1 A in 1 ns, and back to 600 mA after 2 ns, and kept high for 8 ns by assuming the bus speed is half the clock frequency.

Table 7-4. Metal utilization and parasitic values [76]

	Case I: M4 interdigitated	Case II: M4/Substrate noninterdigitated	Case III: M5/M4 interdigitated	Case IV: C4/M4 interdigitated
M5 utilization	—	—	30%/30%	—
M4 utilization	30%/30%	40%/40%	30%/30%	5%/5%
M3 utilization	5%/5%	5%/5%	5%/5%	5%/5%
Number of bond wires	300	300	300	—
Number of bumps	—	—	—	1100
R_{total} (chip and package, mΩ)	20	13	13	2.4
$L_{package}$ (pH)	57	57	50	10.5
L_{chip} (pH)	5	150	2.5	4
L_{total} (pH)	62	200	53	15

Table 7-4 shows the simulation results for the above test conditions in different package and power routing configurations. The metal utilization and approximate resistance and inductance values are summarized in the table. The inductance and metal utilization of the C4 technology is much lower than the cases in the wire-bonding technology.

Table 7-5 shows the result for the V_{cc}–V_{ss} noise comparisons. It is interesting to note that the power saving and peak power conditions cause larger power peaks than the I/O noise. In the past, I/Os have been known to cause most of the noises.

Table 7-5. V_{cc}–V_{ss} performance comparisons [76]

	Case I: M4 interdigitated	Case II: M4/Substrate noninterdigitated	Case III: M5/M4 interdigitated	Case IV: C4/M4 interdigitated
Normal test:				
Average	2.25 V	2.34 V	2.34 V	2.48 V
Minimum	2.24 V	2.32 V	2.33 V	2.46 V
Power saving:				
Minimum	2.23 V	2.29 V	2.30 V	2.41 V
Peak power:				
Average	1.93 V	2.06 V	2.21 V	2.39 V
Minimum	1.84 V	1.83 V	2.08 V	2.29 V
P-to-P noise	0.36 V	0.46 V	0.18 V	0.18 V

The difference now is in the low-voltage swing. The C4 package is clearly the best choice. The C4 undershoots recover within a cycle, so the speed paths, usually in one cycle, are not much affected.

Figure 7-13 shows the V_{cc} and V_{ss} simulated waveforms in the peak power test condition. Case IV in the waveforms is marked as the C4/M4 interdigitated design with the lowest power noise, as in Table 7-5. The peak power test case is described as follows. When the current in the center of the chip suddenly increases by five times, the V_{cc}–V_{ss} is affected significantly in the bond wire cases. Case I drops down and recovers in a couple of clock cycles but the undershoot is not much compared with the final average value. Case II has a significant undershoot.

The high inductance of the substrate tap case causes the center element not to have the benefit of the whole chip's decoupling capacitance. The delay to the edge of the chip also results in a slow start of the currents in the bond wires, so the drop needs about three cycles to settle. Case III is better than Case I, and Case IV with C4 package technology is significantly better. In this case, the settling time is within a clock cycle, and although the minimum V_{cc}–V_{ss} is somewhat affected in the following cycle, the average is about the same.

There is a local decoupling capacitance with time constant R_{dec} · C_{dec}. With a typical L_{chip} of 50–200 pH, including the bond wire and a switching power of maximum 3 A, we can get $R = 2.5$ V/3 A $= 0.83 \ \Omega$ and L/R about 60–240 ps.

With a typical decoupling capacitance of 100 nF and the parasitic added, we can have RC delay = 83 ns. This will clearly dominate over the L/R decoupling. Since the RC time constant is dominant over the L/R time constant, the impedance from the package looks like that of a capacitor. For the typical numbers we have a time constant of 14–28 ns in the bond wire case and less than 10 ns in the C4 case.

As a rule of thumb, one could design the power supply so that C_{dec} takes care of the local drop until the LC can respond. We can increase C_{dec} until this is satisfied. The time constant increases only by the square root of C, whereas the time that the switched power can be sustained by C_{dec} increases linearly with C. With, say, a two times increase in C_{dec}, the improvement in the voltage drop would be 1.41 times.

We have an on-chip decoupling capacitance of 100 nF to take care of local noise. We could deduce the inductance and capaci-

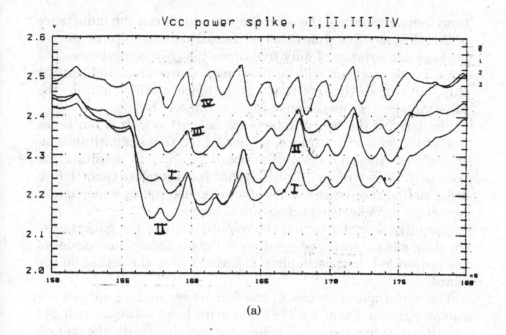

(a)

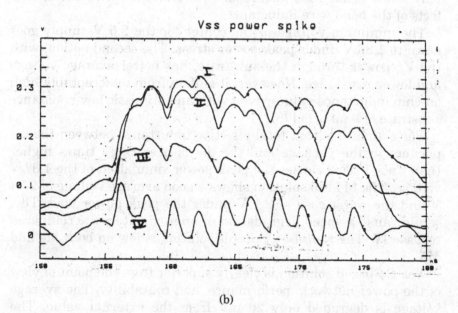

(b)

Figure 7-13. Switching power noises in (a) V_{cc} and (b) V_{ss} for the peak power test [76].

tance requirements of the socket and board. Given the inductance of the pins, say five times that of the package to chip, we need a package capacitance of only five times the chip capacitance. The noise in the package will be five times slower than that on the chip, but due to the five times capacitance it will maintain the same voltage drop as the on-chip capacitance.

Similarly, if the board inductance is 1 nH, which is ten times that of the socket inductance, we need a ten times capacitance increase from the package to the board in order to maintain the voltage at the board for 10 clock cycles. It is therefore clear that in order to find the requirements of on-chip decoupling capacitance, we need a model of the package inductance.

Similarly, in order to find the requirements of the package decoupling capacitance, we need an accurate inductance model of the pin, socket, and decoupling capacitors near the socket on the board.

The performance results of the four power routing options are summarized in Table 7-5 [76]. The wire bond solution with M4 only (Case I) has too much resistance combined with the bad effects of the bond wire inductance.

The minimum voltage in the center for the 2.5 V supply goes down to 1.85 V under peak power stress. The second option, with the V_{ss} power through the substrate, has better average voltage and lower resistance. However, it suffers from unacceptably high on-chip inductance values, even with an optimistic low-resistance substrate of 9 mΩ-cm [76].

The estimate is that total effective inductance between the capacitors in the package and the die is over three times higher than Case I. This causes the peak power simulation of the 1.87 V voltage. The M4/M5 solution shows a good average voltage of 2.34 V and the worst case of 2.1 V under the peak power load. This would cause a speed degradation of approximately 6% compared to Case IV. The routability of this solution is low on both M4 and M5.

The C4-based solution is clearly superior from the point of view of the power network, performance, and routability. The average voltage is degraded only 20 mV from the external value. The worst-case average for a cycle is 2.4 V under the peak power stress.

This gives the best performance of the options considered. The M4 utilization for power is extremely low—only 10%—with the ability to put gaps in the M4 power buses as required for routing. In addition, C4 provides reduction of routing, especially for the I/O areas.

7.4 POWER SUPPLY MEASUREMENT AND VALIDATION

This section will analyze the effectiveness of the on-board decoupling capacitance for microprocessor chips [77]. The model used in the simulation is a PGA-type microprocessor model for the package and the chip parasitic [77].

The main emphasis is to determine the effect that this variation has on the noise seen at the pins and on the die. SPICE circuit simulations, using the frequency domain analysis, were used to assist in the evaluation.

When the number of on-board 1 μF type 1206 decoupling capacitors increases from zero to 35, the resonance frequency increases. Table 7-6 summarizes the optimal measurement bandwidth for each level of board decoupling [77].

The main test points used in these simulations were 0, 1, 5, 10, 20, and 35 decoupling capacitors of the 1 μF ceramic type. The three areas of interest that were examined for all test cases were noise levels at the pins and die and the ratio of correlation between these two parameters. Figure 7-14 shows the noise at the pins in the frequency domain [77].

As the capacitance at the board increases, the resonance frequency also increases. The magnitude of the resonance decreases as the decoupling capacitors are added. Figure 7-15 shows the ra-

Table 7-6. Measurement results for V_{cc}–V_{ss} noises [77]

Number of decoupling capacitors on board	Optimal measured bandwidth (MHz)	Worst-case noise at the die at resonance (V)	Worst-case noise at the pins at resonance (V)
0	45	-1.114	1.048
1	45	0.416	0.364
5	45	0.261	0.180
10	40	0.213	0.116
20	40	0.175	0.066
35	40	0.154	0.040

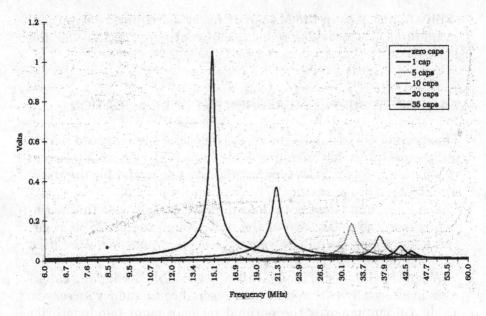

Figure 7-14. Measured noise at the pins of the chip package [77].

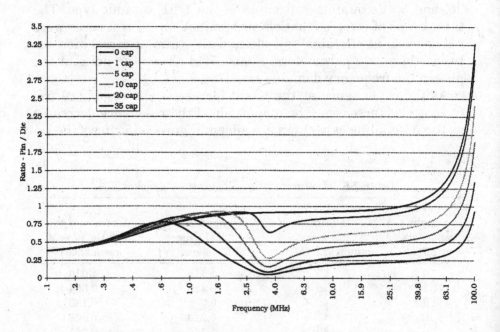

Figure 7-15. Ratio of pins per die versus frequency [77].

tio between the noise at the die and the noise at the pins for various decoupling capacitors at the board.

Figure 7-16 shows the on-board decoupling capacitor model for a ceramic 1 μF capacitor with 15 mΩ parasitic resistance and 2.1 nH parasitic inductance [77].

Figure 7-15 shows that for a majority of the test range frequencies, noise at the die becomes much greater relative to the noise at the pins as the number of capacitors increase on the board. As the frequency increases to greater than 60 MHz, the noise at the pins quickly becomes much greater than the noise at the die, due to the resonance effects. A measurement bandwidth is selected to achieve a more consistent relationship between the pins and die noise over this frequency range.

To understand the effects of varying the number of decoupling capacitors at the board, a model is developed for the package and die's parasitic to be used in the SPICE simulation [77]. Using previously taken test data, it is possible to plot how the simulation model compares to the actual device.

Figure 7-17 shows the discrepancies between the simulation data and empirical data over a range of 20 MHz to 100 MHz for the zero capacitance case. The discrepancy seen between the empirical data and the model is likely due to residual impedance found on the board [77].

A Pentium-II chip scheme dedicated roughly 75% of the M4 layer and 12% of the M3 layer to V_{cc} and V_{ss} routing [78]. The V_{ss} resistance is significantly lower than could have been achieved by using all of M3 and M4 for power routing.

Since M4 is the only thick (low-resistance) metal layer, the main supply current was constrained to the latest M4 routing dimension. Hence, the bulk of the V_{cc} and V_{ss} pins are located to the left and right of the die, where V_{cc} bond wires tie the pack-

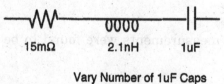

15mΩ   2.1nH   1uF

Vary Number of 1uF Caps
Resistor / # of caps
Inductor / # of caps

Figure 7-16. Modeling of an on-board decoupling capacitor [77].

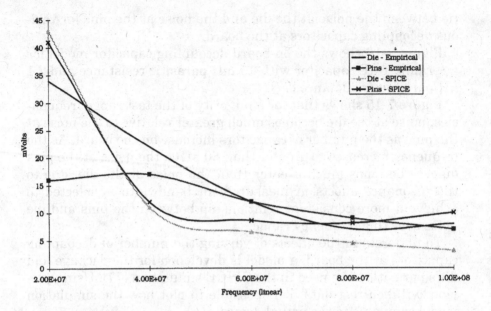

Figure 7-17. Measurement versus simulation data comparison [77].

age V_{cc} planes to the left and right edges of the die, and where a regular array of parallel M4 V_{cc} lines terminate, as shown in Figure 7-18.

The objectives of the measurement are as follows:

1. Feed back measured information into the power grid simulation.
2. Add to the general understanding of the microprocessor power delivery for the preparation of the new process.
3. Determine a solid minimum operating voltage for use in setting the next-step performance goal.
4. Access the impact of adhesive die attachment on the V_{ss} voltage [78].

Two types of measurements were found to be most useful, as follows:

1. DC mapping. A plain wire probe with no transistors or resistors and a voltmeter are used to create a voltage versus position graph of the V_{cc} and V_{ss} planes. This is done for the full

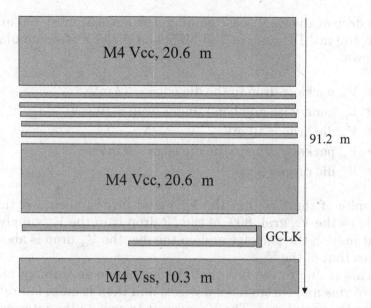

Figure 7-18. M4 power routing patterns [78].

die for a single part, and for a single slice through the die center for many others. The mapping is done with the part running a high-power pattern.

2. AC snapshots. AC waveforms are taken of the V_{cc}, V_{ss}, and V_{cc}-V_{ss} for 33 locations around the die and cavity. Some of these were taken using picoprobes for both transistors and passive differentials, and comparable results were obtained. The differential probes for this type of study are used to automatically subtract the V_{cc} and V_{ss} with a low-noise result. Since there is no need for time-consuming averaging, storing, and subtracting of waveforms, the measurement is an order of magnitude faster than with the FET probes, which allows the engineer more time to search the patterns for the worst-case voltage spikes. Maximum di/dt patterns were used for these snapshots.

The microprocessor performance validation methodology is allowed for a 385 mV drop sustained over multiple cycles, whereas the worst measured drop was only 200 mV. This drop was only sustained over the cycle immediately after a restart of the clocks, and in all other instances lasted less than a phase. The DC voltage drop at the worst-case point on the die is at most 100 mV [78]. The 100 mV DC drop at 133 MHz and 2.755 V is accumulated as follows:

- V_{cc} package drop to the die edge = 32 mV
- V_{cc} bond wire and bond finger drop = 10 mV
- V_{cc} die drop = 40 mV
- V_{ss} package drop to the die edge = 9 mV
- V_{ss} die drop = 9 mV

In spite of the fact that the V_{cc} metal grid is about six times as wide as the V_{ss} grid, 80% of the IR drop is in the V_{cc} supply. Note that in both the package and on the die, the V_{cc} drop is about four times that of the V_{ss}.

This is due to the fact that the V_{cc} has to go through an extra set of vias and through bond wires, and then it must laterally traverse the metal grid. The V_{ss} current travels to the interior of the die on a dedicated metal plane and then has a very short path vertically up through the die.

Had the V_{ss} current been distributed in the same manner as the V_{cc} current, it would have increased the total supply drop by at least 60% and it would have required essentially all of the M4 and M3 planes to be used for the power and clock routing [78]. Thirty-three points around the die and bond cavity were probed using a variety of high-speed probes. All of the locations were probed using the stop-clock pattern, which halts the high power loop and then allows it to resume [78].

Several of the more interesting locations were probed while running patterns tailored to exercise the local power grid, but none of these produced voltage spikes as bad as the stop-clock pattern.

The AC measurements for the patterns and positions include the following:

1. Halt instruction in the high power loop. Probed at the DC hot spot at the bottom center of the die. It takes 200 mV AC transient settling to 100 mV after 20 ns in the stop clock patterns.
2. I/O simultaneous switching pattern. Probed at the middle of the left and right die edges and at the middle of the die, 80

mV transient voltage drop is observed from the V_{ss} grid to the die attachment plane.

3. Simultaneous switching pattern through a large repeater block. Probed at the repeater block, a 2 ns long 180 mV peak glitch is seen in V_{cc}–V_{ss} during switching.

4. Back-side-bus-induced noise from simultaneously switching 28 address lines. A 200 mV peak glitch is observed for less than 2 ns. The worst transient observed in all the probing was observed at the DC hot spot, at the bottom middle of the die, while running the stop-clock pattern. Figure 7-19 shows the 200 mV transient noise in the stop-clock patterns.

Microprobing is very mature, even old fashioned, but we can use it to produce a complete power map, which is extremely valuable. The microprocessor power delivery scheme was quickly proven a success, and the risk taken in running the V_{ss} current through the substrate was shown to be working extremely well for EDA attached parts. The valuable supply voltage information was fed into the design for the B-step, which allows the designers to reset the speed targets [78].

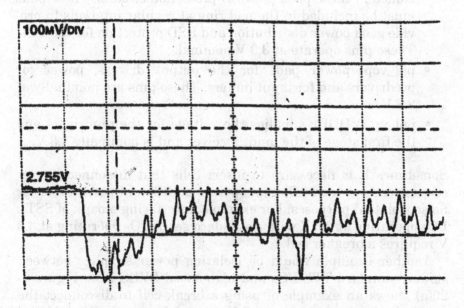

Figure 7-19. V_{cc} transient noise in measurement [78].

7.5   I/O PADS FOR POWER/GROUND SUPPLIES

A set of recommendations can be used for how to place the power and ground pads for the standard I/O library [79]. The following set of I/O power and ground pads are used to supply power and ground pads to the I/O bus structure and internal core in a 0.18 μm process [79].

- pnl_vc (VDD): power pads for core logic and I/O interface with the nominal voltage 1.8 V. These I/O cells provide power to the standard cell core area and interface between the core rings and I/O pads. These pads are paired with the ground pads pnl_gcs.
- pnl_gcs (GND/SGND): ground pads for core standard cell logic, the ground, and substrate ground connections within the I/O set.
- pnl_go (VSSO): ground pads for output drivers. Included in these pads are ESD protection circuits. These pads must be included in the pad ring at regular intervals to provide good power distribution and ESD protection for I/Os.
- pnl_go (VSSO): power pads for the output drivers only. Included in these pads are ESD protection circuits. These pads must be included in the pad ring at regular intervals to provide good power distribution and ESD protection for the I/Os. These pins operate at 3.3 V nominal.
- pnl_vop: power pads for the output drivers, power for predrivers and for input buffers. These pins are nominally at 3.3 V.
- pnl_vp (VDDP): supplies the voltage for the predrivers and the first stage of the input receiver and is nominally 3.3 V.

Sometimes it is necessary to insert cells that disconnect power connections between I/Os that operate at different potential or for noise isolation purposes. For example, interfacing a bank of SSTL I/Os operating at 2.5 V with a standard set of I/Os operating at 3.3 V requires a breaker cell.

Another example would be isolating power supplies between slower, standard TTL I/Os and high-speed LVDS I/Os. Figure 7-20(a) shows an example of using a break cell to disconnect the power and ground between 2.5 V and 3.3 V supplies [79]. Notice that only the power and ground supplies that affect the I/Os are

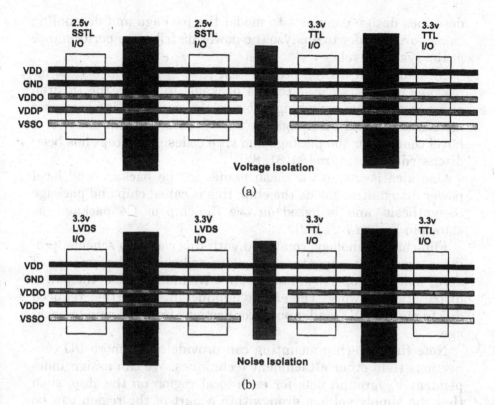

| | 2.5v
SSTL
I/O | | 2.5v
SSTL
I/O | 3.3v
TTL
I/O | | 3.3v
TTL
I/O |

VDD
GND
VDDO
VDDP
VSSO

Voltage Isolation

(a)

| | 3.3v
LVDS
I/O | | 3.3V
LVDS
I/O | 3.3v
TTL
I/O | | 3.3v
TTL
I/O |

VDD
GND
VDDO
VDDP
VSSO

Noise Isolation

(b)

Figure 7-20. Break cells for power supplies [79].

cut. The core power and ground voltages (V_{dd} and V_{cc}) remain intact.

In this diagram, the VDD2.5 supplies power to I/Os that have a 2.5 V reference voltage, and the VDD3.3 supplies power to the I/Os that use a 3.3 V reference voltage. The break cell is inserted between them to separate power and ground for the VDDO, VDDP, and VSSO buses. In Figure 7-20(b), break cells isolate the noise that can occur on power bus connections between high-speed LVDS I/Os and slower TTL I/Os.

7.6 SUMMARY

With the continually increasing clock frequency and performance requirements for chips, the power distribution in both chip and package has to be carefully designed and analyzed. This chapter describes design examples to model the package and decoupling capacitors in order to analyze the power distribution performance at the system level.

In addition, a microprocessor test case is given in details for package options, such as the C4 package, to reduce the *IR* drop and simultaneous switching noise in the design. Microprocessor measurement and I/O design are also discussed in this chapter. A novel concept for the package and chip codesign concept has been discussed in literature [80, 81, 82].

One idea is use to the mesh planes in the package and local power distribution inside the chip; this is called chip and package co-synthesis, and is based on the flip-chip or C4 package, as shown in Figure 7-21 [82].

Flip-chip technology, combined with this codesign scheme, provides significant advantages in noise reduction. The amount of di/dt noise is proportional to the effective inductance of the power distribution network. The effective inductance is further reduced due to multiple V_{dd} and ground connections from the chip to package.

Note that flip-chip mounting can provide many more I/O connections than other attachment techniques. We can assign independent V_{dd}/ground nets for each local region on the chip, such that the supply voltage drop within a part of the region can be small, depending on the partition scales.

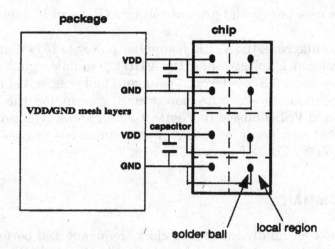

Figure 7-21. Co-design of chip and package power distribution by flip-chip package [82].

GLOSSARY

AC Analysis: transient analysis with detailed voltage distributions over time using stimulus vectors at primary inputs.

Activity-Based Analysis: method used to compute the currents drawn from power supply lines based on the switching activities of circuits.

Back-Annotation: the task of stitching the extracted RC data back to the prelayout circuit netlist to perform the circuit simulation with the interconnect effects.

Block-Level Power Distribution: also called the local power distribution. Connects the power supply from the global power network at the full-chip level and then distributes the power inside a local block region of the chip.

Break Cells: standard cells containing only the power and ground connections. Used to bridge the gaps in cell rows based on standard cell design style.

By-pass Capacitors: the decoupling capacitors added between V_{dd} and V_{ss} networks.

C4 Bump: refers to the solder bump in the special flip-chip package technology. These solder bumps are attached to the top metal of the chip to make the connections to the package.

C4 Package: refers to the flip-chip package technology. Flip-chip connection technology, the first-level chip-to-package connection option, is traditionally regarded as being the controlled collapse chip connection (C4) process. This technology is achieved by distributing the I/O solder bumps over the die, flipping the chips over, aligning them with the contact pads on the substrate, and connecting the solder bumps between the chip and package to make the connections.

Capacitance: the charge storage capability between two conductors.

Capacitance Model: the mathematical equations used to estimate the interconnect capacitances. They contain the variables of geometrical parameters associated with neighboring metal lines.

Chip: the packaged integrated circuits that can be used as a basic building block in complex electrical system designs.

Characterization: the process used to reveal the dependence of electrical performance on design parameters.

Circuit Simulation: the method of using computer programs to model transistors and interconnects, and solve the IV current and voltage equations. The final results are presented as text files or graphical waveforms.

Circuit Timing Analysis: the task of analyzing the hold time and set-up time requirements in sequential logic and other timing constraints across the chip.

Contact: the connection from metal in one layer to diffusion or polysilicon layers.

Core: the chip without the I/O region.

Crossbar Leakage Current: the current from V_{dd} to ground due to a possible short between pMOS and nMOS transistors in the circuit. This is a kind of wasted current for circuit operations.

Decoupling Capacitors: refers to the capacitors added between V_{dd} and V_{ss} lines, used to protect the power supply voltage from sudden switching currents. These decoupling capacitors are required inside the chip, on the package, and in the system board.

Deep-Submicron Process: refers to the VLSI process technology with about 0.18 μm minimum feature size or less.

Delay: the time needed from the 50% V_{dd} of the input signal to the 50% V_{dd} of the output signal through the circuits.

Delta-I Current: same as di/dt noise; the change of switching currents in a short period.

Design Guidelines: the set of guidelines provided by senior designers for the IC design team to follow, to meet the performance, area, and power requirements of chip design.

Design Methodology: the set of design guidelines, CAD tools, and design data flows used to design an IC chip from the conceptual idea to the final working silicon.

Design Rules: the minimum space, minimum width, and minimum coverage, etc. for each physical layer in the VLSI process.

Device-Level Extraction: the modeling of transistors and interconnect RC networks from the physical layout.

Die: refers to the bare chip without the pads and package.

Die Size: the size of the die without the I/O region.

Die Micrograph: photograph of real silicon showing functional blocks and their physical placement in the chip.

DRC: acronym for design rule checking, which verifies any violations in the physical layout against design rules.

Dynamic Analysis: the circuit performance analysis that specifies the input signals for I/Os. Includes dynamic *IR* drop analysis with the input test vectors specified at primary inputs.

ECO (Engineering Change Order): the specifications used to make circuit or logic changes after the initial physical layout has been done. ECO has to be managed carefully since it is time-consuming to change the circuit and layout on a tight schedule.

Electromigration: the phenomenon that causes metal lines to be worn out if the average current density carried by these metal lines exceeds the required upper limit.

ESR: the external series resistance associated with the decoupling capacitance transistor.

Extracted Parasitic: the RC data used to model interconnects in the VLSI physical layout.

Flip-Chip: also called C4 package technology. This technology is achieved by distributing the I/O solder bumps over the die, flipping the chips over, aligning them with the contact pads on the substrate, and connecting the solder bumps between the chip and package to make the connections.

Floor Planning: the arrangement of physical partitions and locations of major functional blocks and I/O pads in the chip. The goal in floor planning is to reduce the total layout area and meet the timing requirements of critical paths.

Full-Chip Power Distribution: the power (V_{dd}, V_{ss}, etc.) networks over the entire chip.

Gate Capacitance: the capacitance from the device gate to the substrate.

Gate Delay: the delay through the logic gate.

GDSII: a binary file used to represent the physical layout. It is in a standard format accepted by most physical design and verification tools.

Global Power Network: the full-chip-level power distribution network on the top metal layers.

Ground Bounce: the variations in the supply voltages, especially in the ground plane, due to the switching of logic gates and output drivers.

Hot Spot: refers to the location in the chip where the local current densities are extremely high due to the high power consumption in this area.

Impedance: the characteristic resistance of the metal traces to the currents carried over them. It is usually characterized in the frequency domain.

Interconnect RC: refers to the resistance and capacitance associated with the metal lines on the routing layers of the chip.

IR Drop: refers to the voltage drop of the power supply current through the resistive network of the distribution network, where the resistive voltage drop is calculated by I (average current) $\cdot R$ (metal resistance) for the power distribution network.

I/O Library: refers to standard cells, particularly those used for I/O functions.

$L \cdot di/dt$ Noise: refers to the voltage drop due to the power supply current change (di) in the Δt rise or fall time through the inductance (L) associated with the power distribution loops. The inductive drop of the power distribution network is calculated by the $L \cdot di/dt$ formula.

Linear Network: refers to the electrical network consisting of the linear elements only, such as the resistor, capacitor, inductor, linear current source, and linear voltage source.

Loop Inductance: the inductance associated with a current loop.

Low-Pass Filter: refers to the RLC network, which filters out the high-frequency harmonics of the input signal and keeps most of the low-frequency harmonics in the output.

LVS: the acronym for layout versus schematic checking, which verifies that the netlist extracted from the physical layout is the same as the prelayout circuit netlist.

Mean Time to Fail (MTF): the average time that a specific product can last, based on thousands of product samples.

Metal Capacitance: the total capacitance from a metal line to adjacent metal lines and to the substrate.

Metal Ions: the atoms and electrons comprising the material of metal layers.

Metal Structures: different combinations of metal lines in the adjacent layout with various metal widths and line-to-line spaces.

Metal Utilization: the ratio of metals used to carry signals compared to the total metals used, including power and ground lines.

Modeling of Power Network: the electrical model of the metal lines and switching currents for the power distribution network.

Noise: refers to the unwanted voltages excited by on-chip activities and those in the signal line or power line.

Noisy Nodes: the nodes in the power distribution network with transient voltages below or above the required voltage thresholds.

On-Chip Inductance: the inductance associated with the on-chip metal lines, especially that of power distribution network.

Pads: the I/O circuits or landing metals from the chip to the outside package.

Parasitics: the interconnect resistance and interconnect capacitance extracted from the physical layout.

Parasitic Capacitance: the capacitance of metal lines in the chip.

Peak Current: the maximum current value over clock cycles.

Physical Design: the task of implementing the logic circuit into the physical layout based on design rules.

Piecewise Linear: A technique in the computer's digitized algorithm for using the linear function in each step size to approximate the general curve function.

Place and Route: to place the standard cells and blocks and then route the cells or blocks based on the circuit netlist. The layout is accomplished using the place and route tool in a standard cell-based design style, especially for the ASIC (application-specific integrated circuit) chip.

Power Bus: the wide power lines used in the power distribution for the chip.

Power Distribution: the task of delivering the power supply (V_{dd}, V_{ss}, etc.) from the power sources to individual transistors on the chip.

Power Grid: the term for the on-chip power distribution network, which is usually routed in the horizontal and vertical grid structure on various metal layers.

Power Grid Analysis: the circuit analysis of voltage drops and electromigration of the power distribution network, based on the power network models for metal lines and device switching currents.

Power Network: refers to the interconnected metal lines on the IC chip used to deliver the V_{dd} and ground voltages.

Power Strap: the wide metal lines used for on-chip power distribution.

Power Supply Voltage: the nominal voltage used by the circuits for correct functions and timing requirements. For example, a 1.8 V supply voltage is usually used in a 0.18 μm process chip.

Power Switching Noise: the voltage variations in the power distribution network due to the switching currents. *IR* drop and $L \cdot di/dt$ drop are the two main sorts of voltage noises causing the power grid to fail.

Prelayout Netlist: the circuit netlist used to specify the layout to be drawn and to describe the connectivity of devices and transistor sizing parameters.

RC Back-Annotation: the process stitching the RC data or RC elements extracted from the VLSI layout back to the prelayout circuit netlist, in order to simulate the circuit with the interconnect RC parasitic.

RC Extraction: the task of modeling metal lines of the chip layout into a distributed RC network together with the transistors.

RC Data: the RC network, extracted from the VLSI physical layout, saved in a file such as the standard parasitic format (SPF) file.

RC Netlist: the netlist used to represent the RC models extracted from the physical layout.

Reflection Noise: the voltage increase or decrease due to the mismatch of the characteristic impedance of the metal line to the load.

Regulator: the circuitry that stabilizes the output power supply to a specified voltage to compensate for supply voltage noise.

Resonance: the phenomenon that results in a cyclic waveform being generated from an RLC network.

RLC Segments: the distributed elements of the resistor, inductor, and capacitor used to model the metal lines of the physical layout, especially for the power distribution grid.

RMS Current: the root mean square current value over clock cycles.

Scaling: refers to the shrinking of the minimal gate length in the IC process by a fixed factor from one generation process to the next. This fixed factor is called the scaling factor. For example, for a 0.18 μm process scaled to a 0.13 μm process, the scaling factor is 0.13/0.18 = 0.72.

Simulation: the method of using computer programs to model transistors and interconnects and solve current and voltage equations.

Simultaneous Switching Noise (SSN): when a number of off-chip loads are switched simultaneously in a digital system, a sudden current change is produced in the power and ground supply networks.

Standard Parasitic Format (SPF) File: the industry-standard file format used to save the RC data extracted from the physical layout of the VLSI chip. It is similar to SPICE netlist format.

Standby Mode: the chip is in a quiet mode with no logic operations.

Static Analysis: circuit performance analysis without the input signals for I/Os; for example, static *IR* drop analysis with no input vectors.

Switching Current: when the logic circuits change the states from logic 1 to logic 0 or from logic 0 to logic 1, a surging current is generated at the source or drain of the transistors, which causes an *IR* drop or *di/dt* noise in the power distribution network.

Switching Factor: a fraction of operating cycles during which the circuit node switches on and off during the clock cycles.

Tap Current: the current source model tied to the power grid used to model the switching currents of transistors.

Technology Parameters: the numbers describing the process technology, such as the minimum gate length, the metal pitches in metal layers, etc. These technology parameters provide the foundation for design rules in IC circuit and layout designs.

Top Metal Layers: the top one or two metal layers of the chip. For example, for a 0.18 μm process chip with six metal layers, Metal 6 and Metal 5 are usually the top metal layers and the global power grid is routed on these top metal layers.

Transistor: the basic device in IC technology used to implement the switching of currents based on the controlling voltages at the terminals. The transistor has four terminals: source, drain, gate, and bulk.

Transistor-Level Simulator: the circuit simulator that uses the transistor device models and interconnect RC or RLC models.

Transmission Line: the long metal trace in the package or board used as an RLC line instead of an RC line.

Unit-Length Capacitance: the capacitance of the metal line per unit length.

Unit-Length Inductance: the inductance of the metal line per unit length.

Unit-Length Resistance: the resistance of the metal line per unit length.

Via: the hole between adjacent metal layers in an IC chip.

Via Resistance: the resistance of each via between two adjacent metal layers.

Vector-Based Analysis: the method used to analyze the switching currents of the circuit, based on the input vectors at chip I/Os.

Voltage Distribution: the various voltage values at the nodes of the power distribution network.

Voltage Fluctuation: the phenomenon caused by supply voltage variations during different time periods and at different locations in the chip.

Voltage Threshold: the upper or lower voltage limits for the supply voltage (V_{dd} or ground) considered as functional for the circuits in the chip.

Voltage Regulation: the step used to adjust the supply voltage to the required stable values. It can be lower or higher than the nominal supply voltage, or even negative for substrate-biasing purposes.

Weak Spot: the location in the power grid where the voltage values are below or above the required voltage thresholds.

Wire Bonding: the chip attachment technology using long lead metals bonded from the package layer to I/O pads.

REFERENCES

1. P. E. Gronowski, W. J. Bowhill, R. P. Preston, M. K. Gowan, and R. L. All-mon, "High-Performance Microprocessor Design," *IEEE Journal of Solid-State Circuits,* Vol. 33, No. 5, May 1998, pp. 676–686.

2. H. H. Chen and D. D. Ling, "Power Supply Noise Analysis Methodology for Deep-Submicron VLSI Chip Design," in *Proceedings of 34th Design Automation Conference,* 1997, pp. 638.

3. A. Dharchoudhury, R. Panda, D. Blaauw, R. Vaidyanathan, B. Tutuianu, and D. Bearden, "Design and Analysis of Power Distribution Networks in PowerPC Microprocessors," in *Proceedings of 35th Design Automation Conference,* 1998, p. 738.

4. G. Steele, D. Overhauser, S. Rochel, S. Z. Hussain, "Full-Chip Verification Methods for DSM Power Distribution Systems," in *Proceedings of 35th Design Automation Conference,* 1998, p. 744.

5. P. C. Li and T. K. Young, "Electromigration: the Time Bomb in Deep-Submicron ICs," *IEEE Spectrum,* Sept. 1996, p. 75.

6. H. B. Bakoglu, *Circuits, Interconnections and Packaging for VLSI,* Addison-Wesley, 1990, Chapter 7.

7. Q. Zhu, "Power Grid Problems and On-Die Decoupling Capacitance Optimization Method," in *Proceedings of IEEE 2nd International Workshop on Chip and Package Co-design,* CPD2000, 2000, p. 46.

8. A. Deutsch et al., "When are Transmission-Line Effects Important for On-Chip Interconnections?," *IEEE Transactions on Microwave Theory and Techniques,* Vol. 45, No. 10, Oct. 1997, p. 1836.

9. N. H. E. Weste and K. Eshraghian, *Principles of CMOS VLSI Design—A System Perspective,* Addison-Wesley, 1992, Chapter 4.

10. M. T. Bohr, "Interconnect Scaling—The Real Limiter to High Performance ULSI," *Solid State Technology Journal,* Sept. 1996, p. 105.

11. A. Odabasioglu, M. Celik, and L. T. Pileggi, "PRIMA: Passive Reduced-Order Interconnect Macromodeling Algorithm," *IEEE Trans. on Computer-*

Aided Design of Integrated Circuits and Systems, Vol. 17, No. 8, Aug. 1998, p. 645.

12. L. T. Pillage and R. A. Rohrer, "Asymptotic Waveform Evaluation for Timing Analysis," *IEEE Trans. on Computer-Aided Design,* Vol. 9, No. 4, April 1990, p. 352.

13. R. Kielkowski, *Inside SPICE,* McGraw-Hill, 1994.

14. K. L. Shepard, S. M. Carey, E. K. Cho, B. W. Curran, R. F. Hatch, D. E. Hoffman, S. A. McCabe, G. A. Northrop, and R. Seigler, "Design Methodology for the S/390 Parallel Enterprise Server G4 Microprocessor," *IBM Journal of Research and Development,* Vol. 41, No. 4/5, May 1997, p. 515.

15. K. L. Shepard and T Zian, "Return-Limited Inductance: A Practical Approach to On-Chip Inductance Extraction," *IEEE Transactions on Computer-Aided Design of Integrated Circuits and Systems,* Vol. 19, No. 4, April 2000, p. 425.

16. M. Basel, "Accurate and Efficient Extraction of Interconnect Circuits for Full-Chip Timing Analysis," in *Proceedings of Design Automation Conference,* 1995, p. 118.

17. A. K. Goel, "High-Speed Interconnections ," Wiley, 1994, Chapter 2.

18. F. Najm, "Transition Density: a New Measure of Activity in Digital Circuits," *IEEE Trans. on Computer-Aided Design,* Vol. 12, No. 2, Feb. 1993, p. 310.

19. M. Xakellis and F. Najm, "Statistical Estimation of the Switching Activity in Digital Circuits," in *Proceedings of 31st ACM/IEEE Design Automation Conference,* 1994, p. 728.

20. E. Grim, Technical Presentations, Intel Corporation, 1999.

21. A. Waizman, Technical Presentations, Intel Corporation, 1998.

22. D. Ayers, Private Communications, Intel Corporation, 1998.

23. T. Burton, Technical Presentations, Intel Corporation, 1998.

24. Q. Zhu, "A New Technique: Decap (Decoupling Capacitance) Sizing and Insertion Based on Power Noise Violation Nodes," *USA Patent # 6446016,* Sep. 2002.

25. Y. L. Le Coz and R. B. Iverson, "A Stochastic Algorithm for High-Speed Capacitance Extraction in Integrated Circuits," *Solid-State Electronics,* Vol. 35, No. 7, July 1992, p. 1005.

26. P. Larsson, "Resonance and Damping in CMOS Circuits with on-Chip Decoupling Capacitance," *IEEE Transactions on Circuits and Systems—I: Fundamental Theory and Applications,* Vol. 45, No. 8, Aug. 1998, p. 849.

27. The National Technology Roadmap for Semiconductors, Semiconductor Research Corporation, 1997.

28. Q. Zhu and W. W.-M. Dai, "High Speed Clock Network Sizing Optimization Based on Distributed RC and RLC Interconnect Models," *IEEE Transactions on Computer-Aided Design of Integrated Circuits and Systems,* Vol. 15, No. 9, Sept. 1996, p. 1106.

29. G. K. Rao, *Multilevel Interconnect Technology,* McGraw-Hill, 1993.

30. TSMC Technology Roadmap, www.tsmc.com, 2002.

31. TSMC 0.18 μm Logic 1P6M Salicide 1.5 V/3.3 V Design Rule, *Taiwan Semiconductor Manufacturing Co.,* Nov. 2000.

32. P. E. Allen and D. R. Holberg, *CMOS Analog Circuit Design,* Oxford University Press, 1987.

33. *Star-RCXT User Guide,* Synopsys Corporation, 2002.

34. *Star-SimXT User Guide,* Synopsys Corporation, 2002.

35. R. Kumar, *Noise Design and Analysis,* Intel Corporation, 1997.

36. A. K. Goel, *High-Speed VLSI Interconnections: Modeling, Analysis and Simulation,* Wiley, 1994.

37. P. DeWilde and Z.-Q. Ning, *Models for Large Integrated Circuits,* Kluwer Academic Publishers, 1990.

38. C. S. Walker, *Capacitance, Inductance, and Crosstalk Analysis,* Artech House, 1990.

39. *Virtuoso User Guide,* Cadence Design Systems, Inc., 2002.

40. *HSPICE User Guide,* Synopsys, Inc., 2002.

41. N. D. Arora et al., "Modelling and Extraction of Interconnect Capacitances for Multilayer VLSI Circuits," *IEEE Trans. on Computer-Aided Design of Integrated Circuits and Systems,* Vol. 15, No. 1, pp. 58–67, Jan. 1996.

42. Q. Zhu, "Star-RCXT Capacitance Accuracy Study," T-RAM, Inc., Feb. 2002.

43. J. Savoj and B. Razavi, *High-Speed CMOS Circuits for Optical Receivers,* Kluwer Academic Publishers, 2001.

44. W. S. Song and L. A. Glasser, "Power Distribution Techniques for VLSI Circuits," *IEEE Journal of Solid-State Circuits,* Vol. SC-21, No. 1, Feb. 1986.

45. Q. Zhu, *Power Grid Design and Specifications,* Chameleon Systems, Inc., 2001.

46. D. Ayers, *Microprocessor Power Network Design,* Intel Corporation, 1998.

47. Y. Jiang, *P6C AC Analysis,* Intel Corporation, 1995.

48. Y. Jiang, *P6C Decoupling Capacitor Methodology,* Intel Corporation, 1995.

49. H. H. Chen and D. D. Ling, "Power Supply Analysis Methodology for Deep-Submicron VLSI Chip Design," in *Proceedings of 34th Design Automation Conference,* 1997, p. 638.

50. B. J. Rubin, "An Electromagnetic Approach for Modeling High-Performance Computer Package," *IBM Journal of Research and Development,* Vol. 34, pp. 585–600, July 1990.

51. *VoltageStorm Transistor-Level PGS User Guide,* Cadence Design Systems, Inc., 2002.

52. A. Chandrakasan, W. J. Bowhill, and F. Fox, *Design of High-Performance Microprocessor Circuits,* IEEE Press, 2001, Chapter 24.

53. Q. Zhu and J. Pabustan, *Post-Layout Static IR Analysis Flow Based on Simplex Tool,* Chameleon Systems, Inc., 2001.

54. V. L. Bars, *IR Drop Evaluation in a Power/Ground Mesh,* Project Report on UC Santa Cruz Extension, 1997.

55. S. Chowdhury and J.S. Barkatullah, "Estimation of Maximum Currents in

MOS IC Logic Circuits," *IEEE Transactions on Computer-Aided Design*, Vol. 9, No. 6, June 1990, pp. 642–654.

56. J. N. Kozhaya and F. N. Najm, "Power Estimation for Large Sequential Circuits," *IEEE Transactions on VLSI Systems*, Vol. 9, No. 2, April 2001, pp. 400–407.

57. M. S. Hsiao, E. M. Rudnick, and J. H. Patel, "Peak Power Estimation of VLSI Circuits: New Peak Power Measures," *IEEE Transactions on VLSI Systems*, Vol. 8, No. 4, August 2000, pp. 435–439.

58. Y.-M. Jiang, K.-T. Cheng, and A. Krstic, "Estimation of Maximum Power and Instantaneous Current Using a Genetic Algorithm," in *Proceedings of Custom Integrated Circuits Conference*, 1997, pp. 135–138.

59. *Star-RCXT User Guide*, Synopsys Inc., 2001.

60. *Fire & Ice User Guide*, Cadence Design Systems, Inc., 2001.

61. T. Mozdzen, J. Barkatullah, S. Rajgopal, and D. Weiss, *Management of Power Supply Noise Using Die, Package and Board Level Solutions*, Intel Corporation, 1995.

62. A. Dharchoudhury, R. Panda, D. Blaauw, R. Vaidyanathan, B. Tutuianu, and D. Bearden, "Design and Analysis of Power Distribution Networks in PowerPC Microprocessors," in *Proceedings of 35th ACM/IEEE Design Automation Conference*, 1998, pp. 738–743.

63. P. E. Gronowski, W. J. Bowhill, R. P. Preston, M. K. Gowan, and R. L. Allmon, "High-Performance Microprocessor Design," *IEEE Journal of Solid-State Circuits*, Vol. 33, No. 5, May 1998, pp. 676–686.

64. *HiP7 Design Manual*, Motorola, Inc., Oct. 2002.

65. *ElectronStorm Manual 3.1*, Cadence Design Systems, Inc., 2001.

66. R. Senthinathan, S. Fischer, H. Rangchi and H. Yazdanmehr, "A 650-MHz, IA-32 Microprocessor with Enhanced Data Streaming for Graphics and Video," *IEEE Journal of Solid-State Circuits*, Vol. 34, No. 11, Nov. 1999, pp. 1454–1465.

67. G. K. Konstadinidis, K. Normoyle, S. Wong, S. Bhutani, H. Stuimer, T. Johnson, A. Smith, D. Y. Cheung, F. Romano, S. Yu, S.-H. Oh, V. Melamed, S. Narayanan, D. Bunsey, C. Khieu, K. J. Wu, R. Schmitt, A. Dumlao, M. Sutera, J. Chau, K. J. Lin and W. S. Coates, "Implementation of a Third-Generation 1.1-GHz 64-bit Microprocessor," *IEEE Journal of Solid-State Circuits*, Vol. 37, No. 11, Nov. 2002, pp. 1461–1469.

68. H. Mizuno, K. Ishibashi, T. Shimura, T. Hattori, S. Narita, K. Shiozawa, S. Ikeda, and K. Uchiyama, "An 18-µA Standby-Current 1.8 V 200 MHz Microprocessor with Self Substrate-Biased Data Retention Mode," *IEEE Journal of Solid-State Circuits*, Vol. 34, No. 11, Nov. 1999, pp. 1492–1500.

69. C. F. Webb et al., "A 400-MHz S/390 Microprocessor," *IEEE Journal of Solid-State Circuits*, Vol. 32, No. 11, Nov. 1997, pp. 1665–1675.

70. R. Heald et al., "A Third-Generation SPARC V9 64-b Microprocessor," *IEEE Journal of Solid-State Circuits*, Vol. 35, No. 11, Nov. 2000, pp. 1526–1538.

71. S. Rusu and G. Singer, "The First IA-64 Microprocessor," *IEEE Journal of Solid-State Circuits*, Vol. 35, No. 11, Nov. 2000, pp. 1539–1544.

72. C. Nicoletta et al., "A 450-MHz RISC Microprocessor with Enhanced In-

struction Set and Copper Interconnect," *IEEE Journal of Solid-State Circuits*, Vol. 34, No. 11, Nov. 1999, pp. 1478–1491.

73. C. C. Wong, "Flip Chip Connection Technology," in *Multichip Module Technologies and Alternatives: The Basics*, Edited by D. A. Doane and P. D. Franzon, Van Nostrand Reinhold, 1993.

74. P. D. Franzon, "Electrical Design of Digital Multichip Module," in *Multichip Module Technologies and Alternatives: The Basics*, Edited by D. A. Doane and P. D. Franzon, Van Nostrand Reinhold, 1993.

75. A. Chakrabarti, *A Preliminary Analysis of Decoupling in Package and Chips*, LSI Logic Corp., 1994.

76. B. Kleveland and J. Prak, *Chip and Package Power Supply Analysis*, Intel Corporation, 1993.

77. B. Jocobs, *VCC/VSS Noise Measurement Bandwidth vs. Motherboard Decoupling*, Intel Corporation, 1995.

78. T. Burton, *Power Grid Measurement Show Big Performance Win*, Intel Corporation, 1996.

79. *Application Note: Recommended Placement for Power and Ground Pads for the Standard I/O Pad Library*, Nurlogic Design, Inc., 2001.

80. Q. Zhu and S. Tam, "Package Clock Distribution Design Optimization for High-Speed and Low-Power VLSIs," *IEEE Transactions on CPMT/Advanced Packaging*, Vol. 20, No. 1, pp. 56–63, Feb. 1997.

81. Q. Zhu and W. W.-M. Dai, "Planar Clock Routing for Chip and Package Co-Design," *IEEE Transactions on VLSI Systems*, Vol. 4, No. 2, pp. 210–226, June 1996.

82. Q. Zhu, *Chip and Package Co-design of Clock Networks*, Ph.D. Thesis, University of California, Santa Cruz, June 1995.

83. Q. Zhu, "An On-Chip Decoupling Capacitance Allocation Method," in *Proceedings of Northeast Workshop on Circuits and Systems*, Canada, May 2003, pp. 121–124.

INDEX